# A Brief Atlas of the
# Human
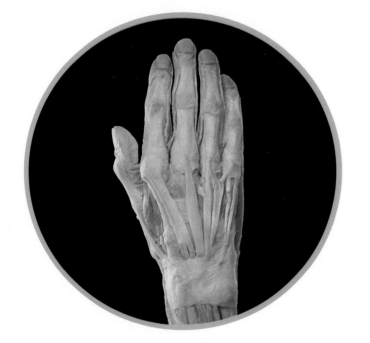
# Body

Matt Hutchinson
Jon Mallatt
Elaine N. Marieb

Photographs by Ralph T. Hutchings

Benjamin
Cummings

SAN FRANCISCO  BOSTON  NEW YORK
CAPE TOWN  HONG KONG  LONDON  MADRID  MEXICO CITY
MONTREAL  MUNICH  PARIS  SINGAPORE  SYDNEY  TOKYO  TORONTO

Publisher: Daryl Fox
Managing Editor, Editorial: Kay Ueno
Managing Editor, Production: Wendy Earl
Production Manager: Janet Vail
Production Editor/Photo Coordinator: Kelly Murphy
Photographer: Ralph T. Hutchings*
Cover and Front Matter Designer: Lillian Carr
Compositor: Precision Graphics

*Figure 46 from *The Bassett Atlas of Human Anatomy* by Robert A. Chase, ©1989, The Benjamin/Cummings Publishing Company, Inc.
Figure 53a photographed by John Martinek, Kirkwood Community College.

ISBN 0-8053-5336-4
1 2 3 4 5 6 7 8 9 10—VHP—06 05 04 03
www.aw.com/bc

Benjamin
Cummings

# Preface

*A Brief Atlas of the Human Body* evolved from a joint inspiration of Elaine N. Marieb and Benjamin Cummings Publisher Daryl Fox: *The Atlas of the Human Skeleton,* which was packaged with *Human Anatomy & Physiology,* Fifth Edition, and *Human Anatomy,* Third Edition, upon their publication in 2000. The full-color views of the human skeleton in that atlas provided students with a degree of clarity and scale that could never be achieved within a textbook alone.

Building upon our original vision, this new atlas features 47 soft tissue images, in addition to the 104 bone images featured in the *Atlas of the Human Skeleton.* Elaine Marieb chose the soft tissue views. Matt Hutchinson, of Washington State University, took on the arduous task of labeling each structure. Jon Mallatt, co-author of the human anatomy text, scrutinized and approved the views, leaders, and labels. References to related atlas images herein can be found in the illustration figure legends of both the anatomy and the anatomy & physiology textbooks.

Ralph T. Hutchings, formerly with The Royal College of Surgeons of England, photographed each of the bone structures found in this book, and most of the soft tissue images. His reputation as an anatomical photographer preceded him, and we certainly were not disappointed — the quality of his work is here for all to see. We are most grateful to him for lending his expertise to this project, and for his good humor and ready willingness to meet our demands. We are grateful to John Martinek of Kirkwood Community College, who contributed his excellent photograph of the internal surface of the stomach (Figure 53a).

We are hopeful that *A Brief Atlas of the Human Body* proves to be as relevant to students and instructors as its predecessor. Benjamin Cummings would welcome your comments and suggestions, which may be sent to the following address:

PUBLISHER
APPLIED SCIENCES
BENJAMIN CUMMINGS
1301 SANSOME STREET
SAN FRANCISCO, CA 94111

# Contents

## Part I  Bones of the Human Skeleton

## Part II    Soft Tissue of the Human Body

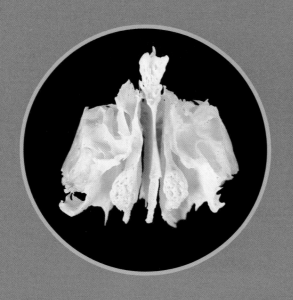

# Part I

# Bones of the
# Human Skeleton

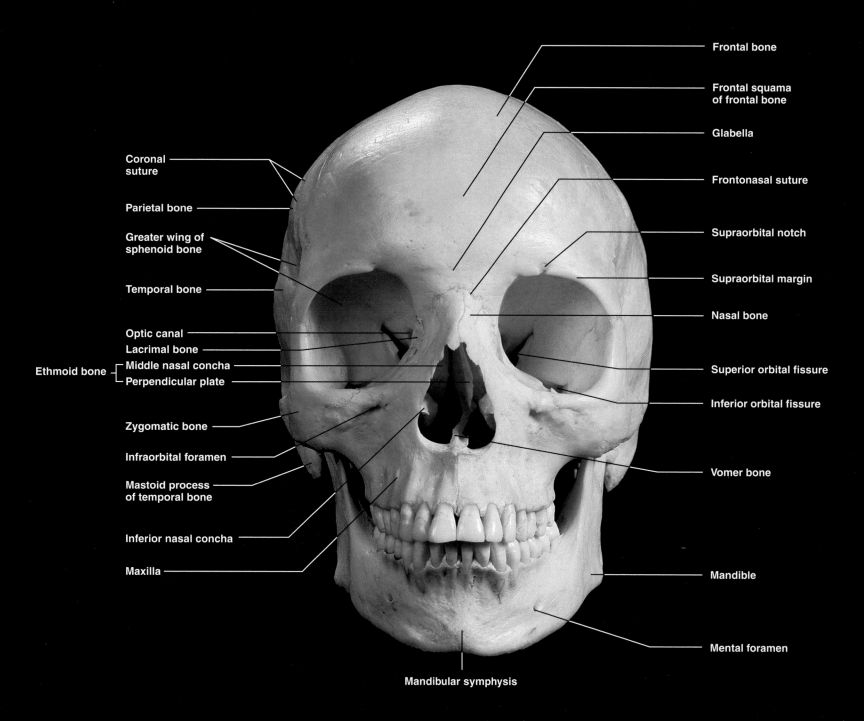

Coronal suture

Parietal bone

Greater wing of sphenoid bone

Temporal bone

Optic canal

Lacrimal bone

Ethmoid bone ⌈ Middle nasal concha
              ⌊ Perpendicular plate

Zygomatic bone

Infraorbital foramen

Mastoid process of temporal bone

Inferior nasal concha

Maxilla

Frontal bone

Frontal squama of frontal bone

Glabella

Frontonasal suture

Supraorbital notch

Supraorbital margin

Nasal bone

Superior orbital fissure

Inferior orbital fissure

Vomer bone

Mandible

Mental foramen

Mandibular symphysis

**Figure 1**   Skull, anterior view.

1

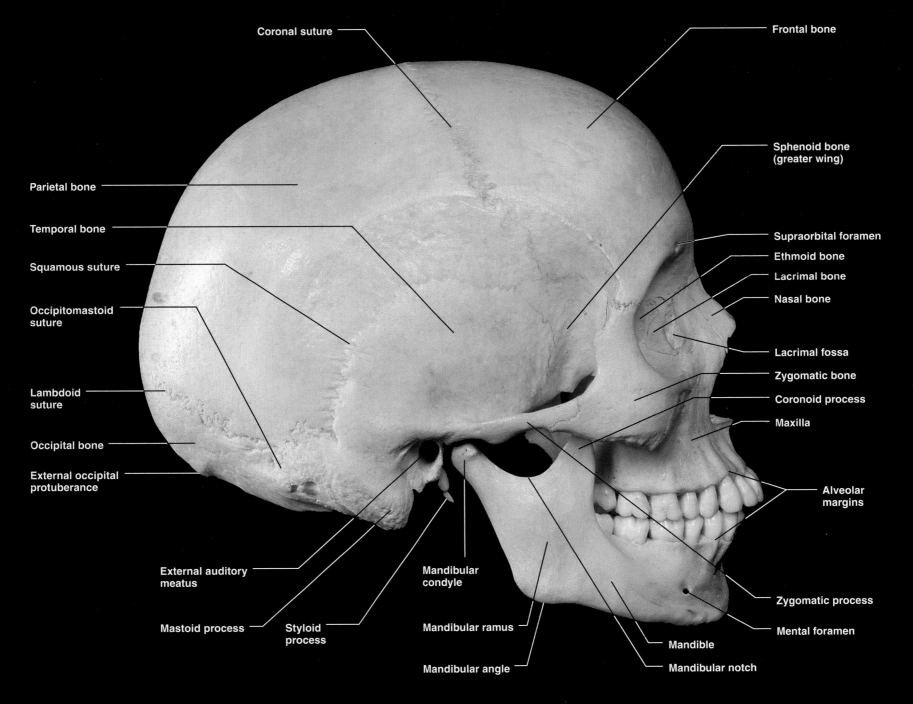

**Figure 2** Skull, right external view of lateral surface.

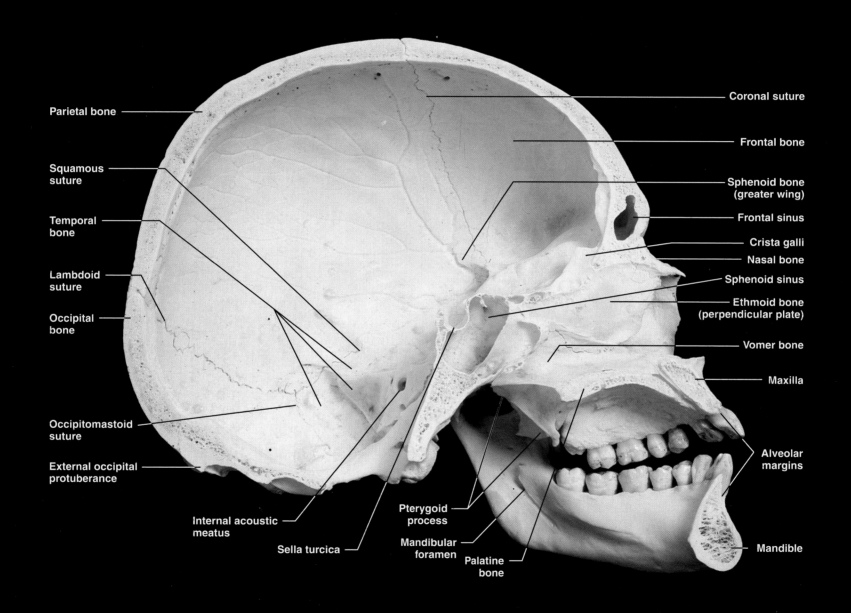

**Figure 3** Skull, internal view of left lateral aspect.

Parietal bone

Squamous suture

Temporal bone

Lambdoid suture

Occipital bone

Occipitomastoid suture

External occipital protuberance

Internal acoustic meatus

Sella turcica

Pterygoid process

Mandibular foramen

Palatine bone

Coronal suture

Frontal bone

Sphenoid bone (greater wing)

Frontal sinus

Crista galli

Nasal bone

Sphenoid sinus

Ethmoid bone (perpendicular plate)

Vomer bone

Maxilla

Alveolar margins

Mandible

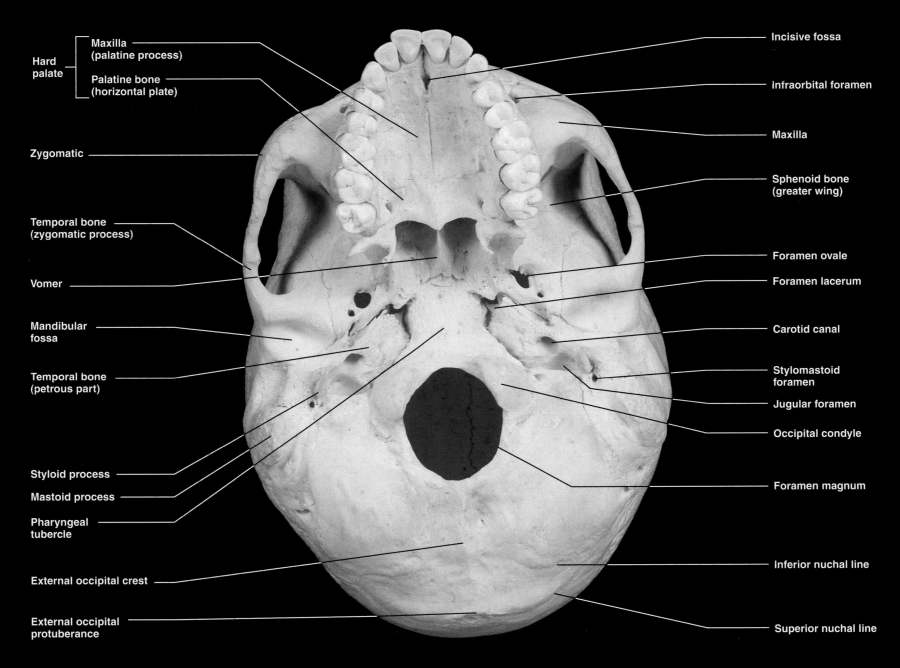

**4**

**Figure 4** Skull, external view of base.

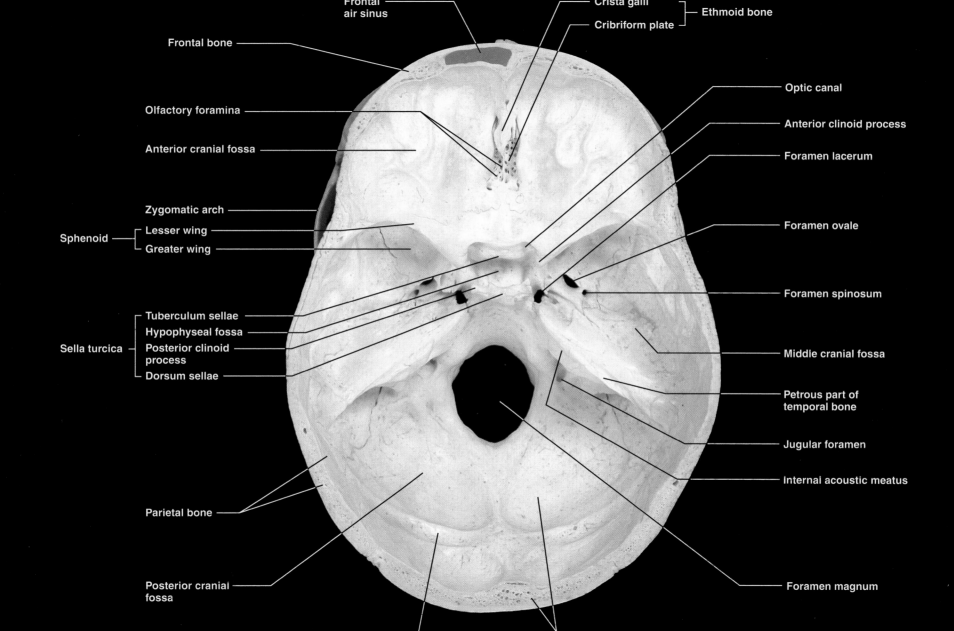

Frontal
air sinus

Crista galli

Cribriform plate

Ethmoid bone

Frontal bone

Optic canal

Olfactory foramina

Anterior clinoid process

Anterior cranial fossa

Foramen lacerum

Zygomatic arch

Foramen ovale

Sphenoid — Lesser wing

Greater wing

Foramen spinosum

Tuberculum sellae

Middle cranial fossa

Hypophyseal fossa

Sella turcica — Posterior clinoid
process

Petrous part of
temporal bone

Dorsum sellae

Jugular foramen

Internal acoustic meatus

Parietal bone

Foramen magnum

Posterior cranial
fossa

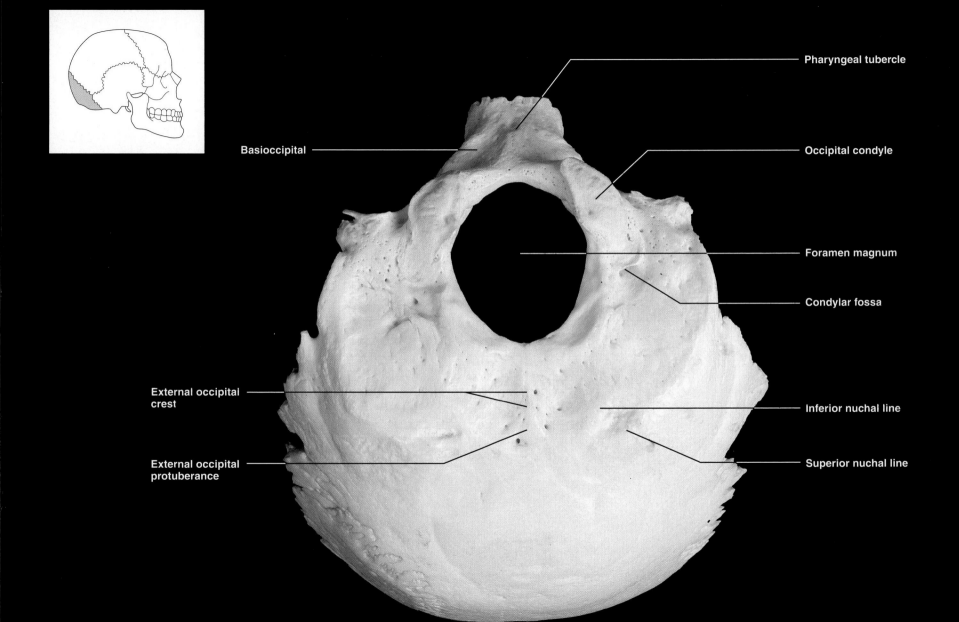

Pharyngeal tubercle

Basioccipital

Occipital condyle

Foramen magnum

Condylar fossa

External occipital crest

Inferior nuchal line

External occipital protuberance

Superior nuchal line

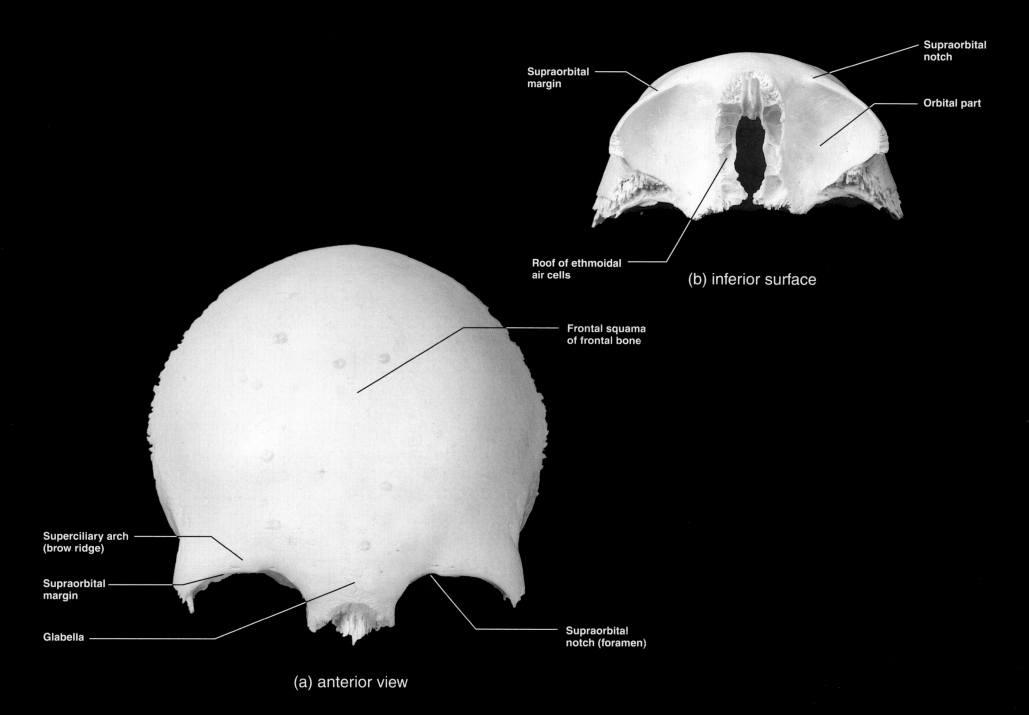

**Figure 7** Frontal bone.

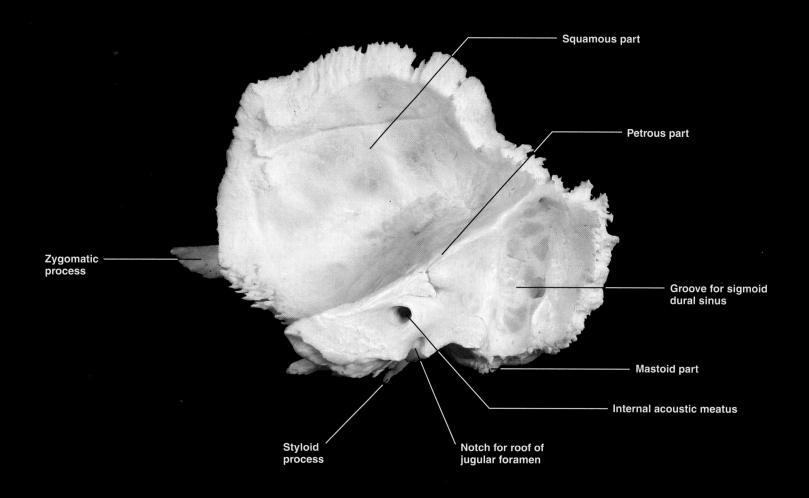

Squamous part

Petrous part

Zygomatic process

Groove for sigmoid dural sinus

Mastoid part

Internal acoustic meatus

Styloid process

Notch for roof of jugular foramen

(b) right medial view

**Figure 8** Temporal bone.

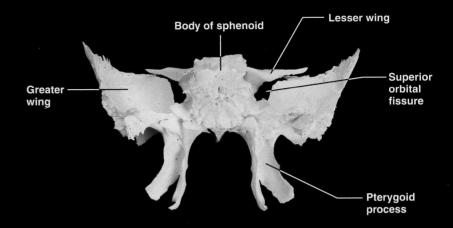

Body of sphenoid

Lesser wing

Greater wing

Superior orbital fissure

Pterygoid process

(b) posterior view

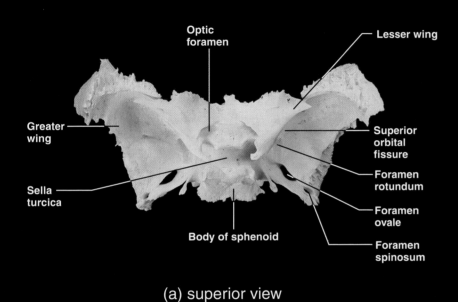

Optic foramen

Lesser wing

Greater wing

Superior orbital fissure

Foramen rotundum

Sella turcica

Foramen ovale

Body of sphenoid

Foramen spinosum

(a) superior view

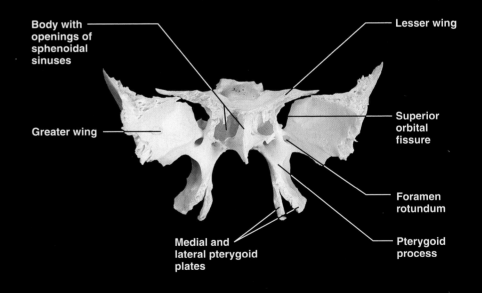

Body with openings of sphenoidal sinuses

Lesser wing

Greater wing

Superior orbital fissure

Foramen rotundum

Medial and lateral pterygoid plates

Pterygoid process

(c) anterior view

**Figure 9** Sphenoid bone.

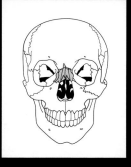

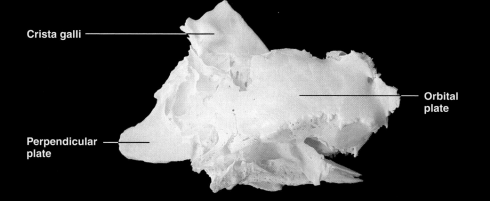

Crista galli

Orbital plate

Perpendicular plate

(a) left lateral surface

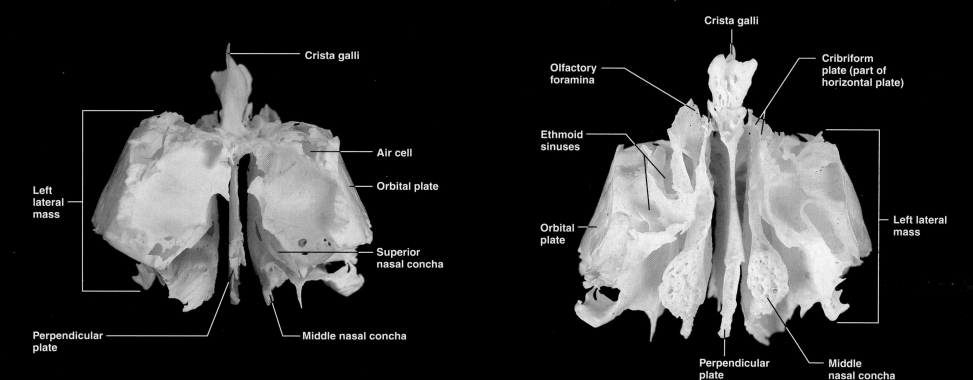

Crista galli

Air cell

Orbital plate

Superior nasal concha

Left lateral mass

Perpendicular plate

Middle nasal concha

Crista galli

Olfactory foramina

Cribriform plate (part of horizontal plate)

Ethmoid sinuses

Orbital plate

Left lateral mass

Perpendicular plate

Middle nasal concha

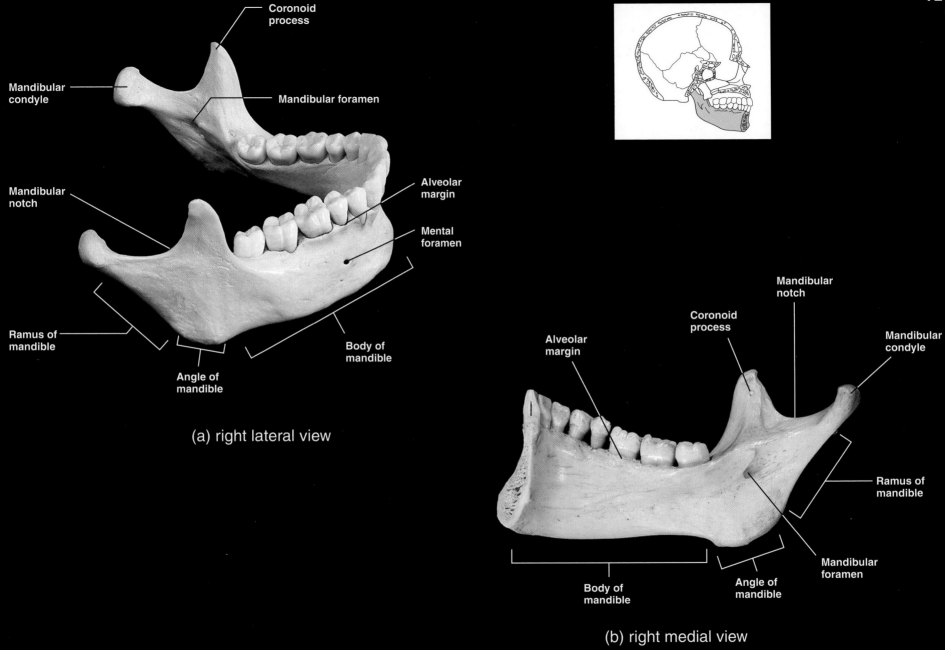

Coronoid process

Mandibular condyle

Mandibular foramen

Mandibular notch

Alveolar margin

Mental foramen

Ramus of mandible

Angle of mandible

Body of mandible

(a) right lateral view

Alveolar margin

Coronoid process

Mandibular notch

Mandibular condyle

Ramus of mandible

Body of mandible

Angle of mandible

Mandibular foramen

(b) right medial view

**Figure 11**  Mandible.

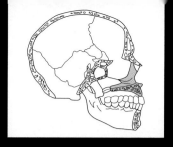

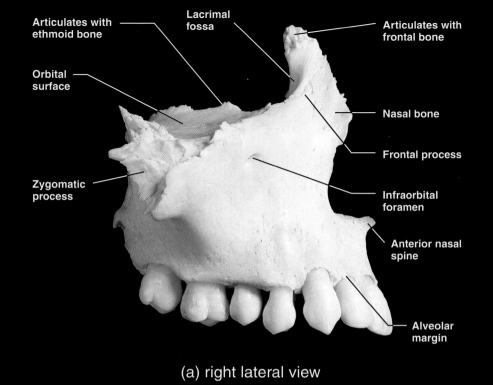

Articulates with
ethmoid bone

Lacrimal
fossa

Articulates with
frontal bone

Orbital
surface

Nasal bone

Frontal process

Zygomatic
process

Infraorbital
foramen

Anterior nasal
spine

Alveolar
margin

**(a) right lateral view**

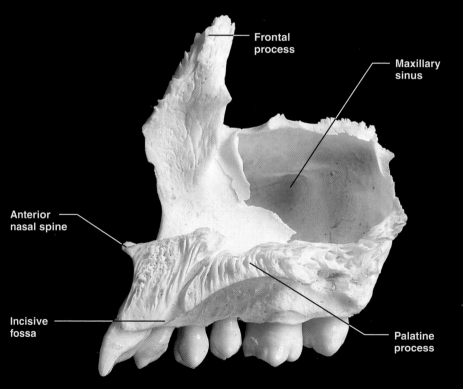

Frontal
process

Maxillary
sinus

Anterior
nasal spine

Incisive
fossa

Palatine
process

**(b) right medial view**

**Figure 12** Maxilla.

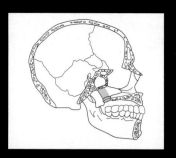

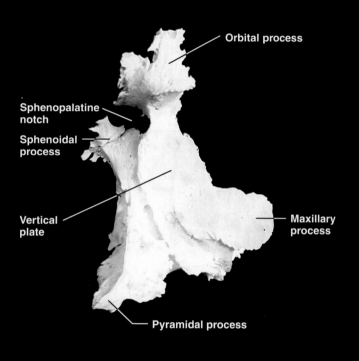

Orbital process

Sphenopalatine notch

Sphenoidal process

Vertical plate

Maxillary process

Pyramidal process

(a) right lateral view

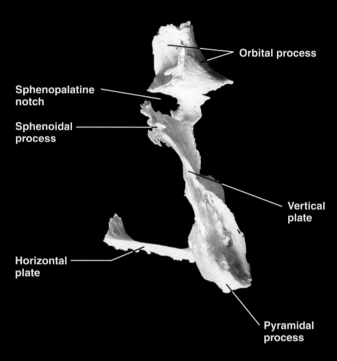

Orbital process

Sphenopalatine notch

Sphenoidal process

Vertical plate

Horizontal plate

Pyramidal process

(b) right posterior view

**Figure 13**   Palatine bone.

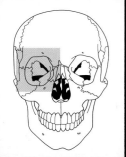

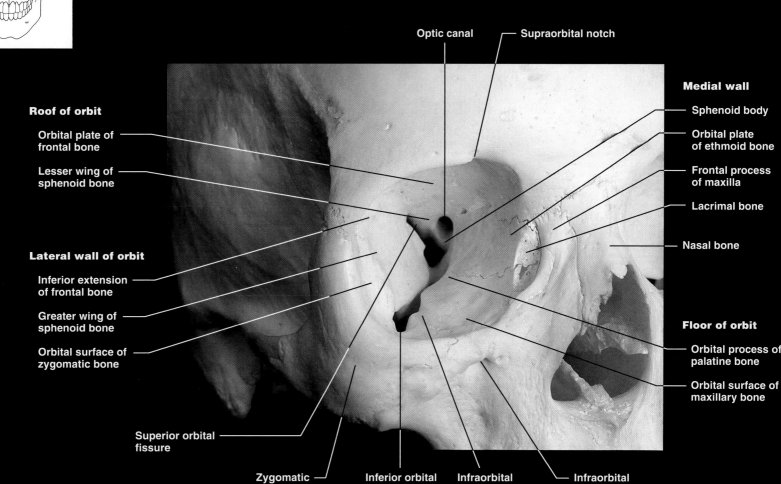

Optic canal — Supraorbital notch

**Medial wall**

**Roof of orbit**

Orbital plate of
frontal bone

Lesser wing of
sphenoid bone

Sphenoid body

Orbital plate
of ethmoid bone

Frontal process
of maxilla

Lacrimal bone

Nasal bone

**Lateral wall of orbit**

Inferior extension
of frontal bone

Greater wing of
sphenoid bone

Orbital surface of
zygomatic bone

**Floor of orbit**

Orbital process of
palatine bone

Orbital surface of
maxillary bone

Superior orbital
fissure

Zygomatic — Inferior orbital — Infraorbital — Infraorbital

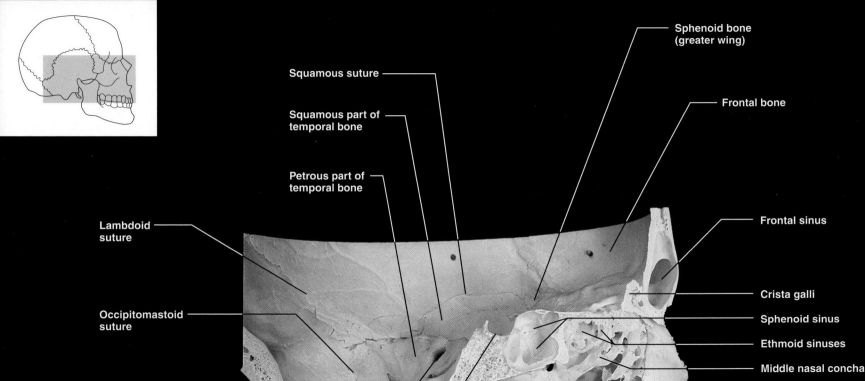

Squamous suture

Squamous part of
temporal bone

Petrous part of
temporal bone

Lambdoid
suture

Occipitomastoid
suture

Sphenoid bone
(greater wing)

Frontal bone

Frontal sinus

Crista galli

Sphenoid sinus

Ethmoid sinuses

Middle nasal concha

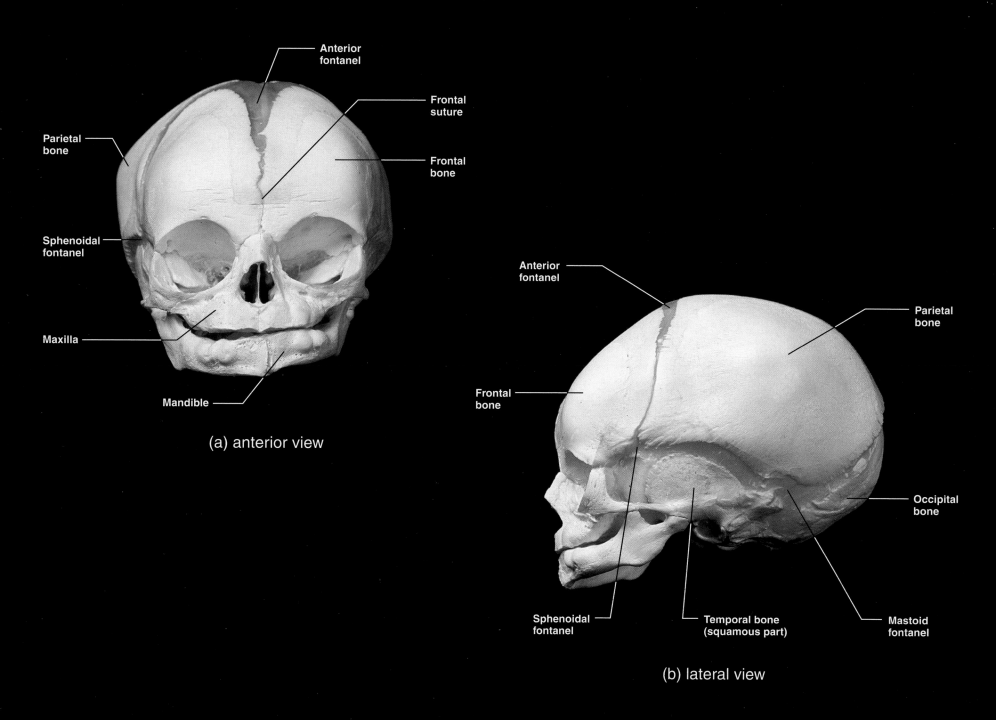

**Figure 16**  Fetal skull.

(a) anterior view

Anterior fontanel

Frontal suture

Parietal bone

Frontal bone

Sphenoidal fontanel

Maxilla

Mandible

(b) lateral view

Anterior fontanel

Parietal bone

Frontal bone

Occipital bone

Sphenoidal fontanel

Temporal bone (squamous part)

Mastoid fontanel

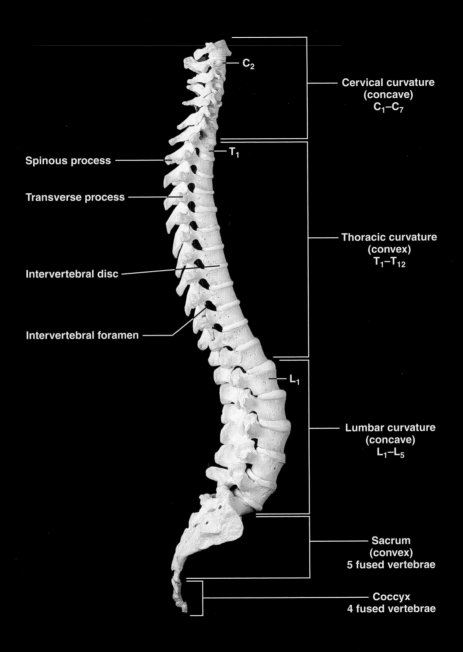

$C_2$

Cervical curvature
(concave)
$C_1$–$C_7$

$T_1$

Spinous process

Transverse process

Thoracic curvature
(convex)
$T_1$–$T_{12}$

Intervertebral disc

Intervertebral foramen

$L_1$

Lumbar curvature
(concave)
$L_1$–$L_5$

Sacrum
(convex)
5 fused vertebrae

Coccyx
4 fused vertebrae

(a) right lateral view

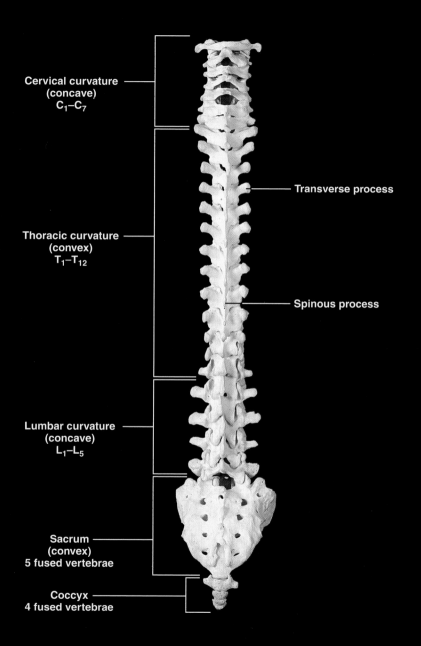

Cervical curvature
(concave)
$C_1-C_7$

Thoracic curvature
(convex)
$T_1-T_{12}$

Transverse process

Spinous process

Lumbar curvature
(concave)
$L_1-L_5$

Sacrum
(convex)
5 fused vertebrae

Coccyx
4 fused vertebrae

(b) posterior view

**Figure 17** Articulated vertebral column.

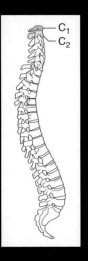

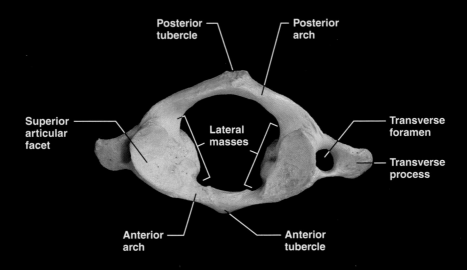

Posterior tubercle

Posterior arch

Superior articular facet

Lateral masses

Transverse foramen

Transverse process

Anterior arch

Anterior tubercle

(a) atlas, superior view

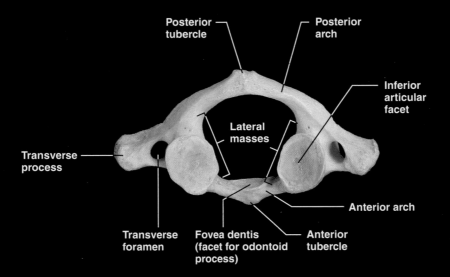

Posterior tubercle

Posterior arch

Inferior articular facet

Lateral masses

Transverse process

Transverse foramen

Fovea dentis (facet for odontoid process)

Anterior arch

Anterior tubercle

(b) atlas, inferior view

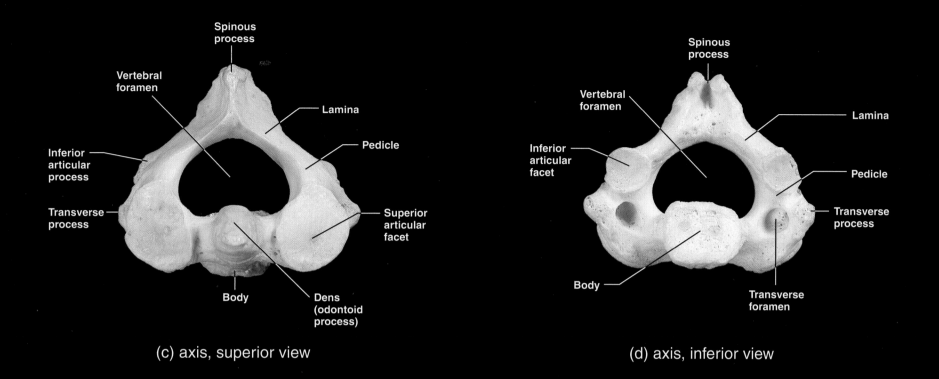

(c) axis, superior view

(d) axis, inferior view

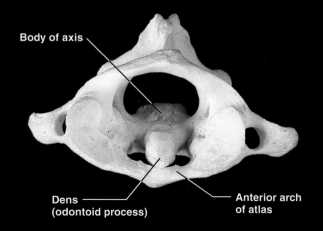

(e) articulated atlas and axis, superior view

**Figure 18** Various views of vertebrae C$_1$ and C$_2$.

21

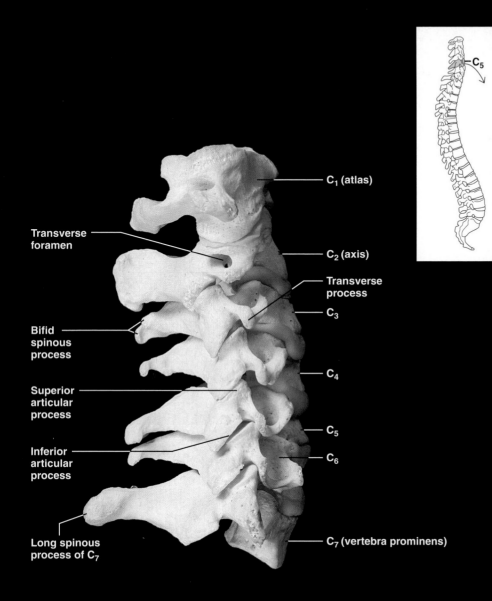

C₁ (atlas)

Transverse foramen

C₂ (axis)

Transverse process

C₃

Bifid spinous process

C₄

Superior articular process

C₅

Inferior articular process

C₆

Long spinous process of C₇

C₇ (vertebra prominens)

(a) right lateral view of articulated cervical vertebrae

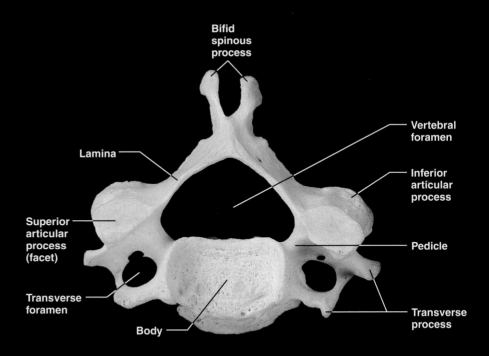

Bifid spinous process

Vertebral foramen

Lamina

Inferior articular process

Superior articular process (facet)

Pedicle

Transverse foramen

Body

Transverse process

(b) fifth (typical) cervical vertebra, superior view

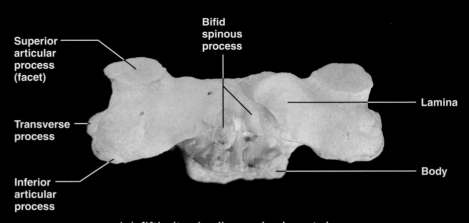

Superior articular process (facet)

Bifid spinous process

Transverse process

Lamina

Inferior articular process

Body

(c) fifth (typical) cervical vertebra, posterior view

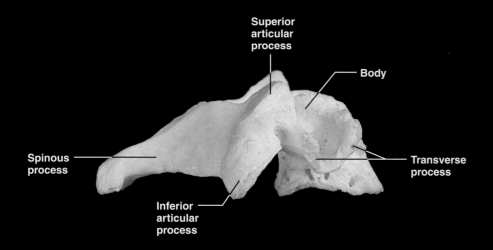

**Superior articular process**

**Body**

**Spinous process**

**Transverse process**

**Inferior articular process**

(d) fifth (typical) cervical vertebra, right lateral view

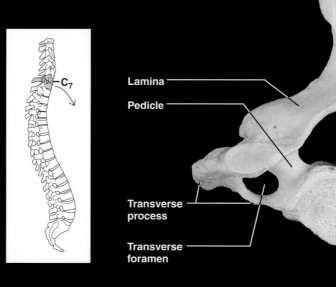

C₇

**Spinous process**

**Vertebral foramen**

**Lamina**

**Pedicle**

**Inferior articular process**

**Transverse process**

**Transverse foramen**

**Superior articular process (facet)**

**Body**

(e) vertebra prominens (C₇), superior view

**Figure 19**  Cervical vertebrae.

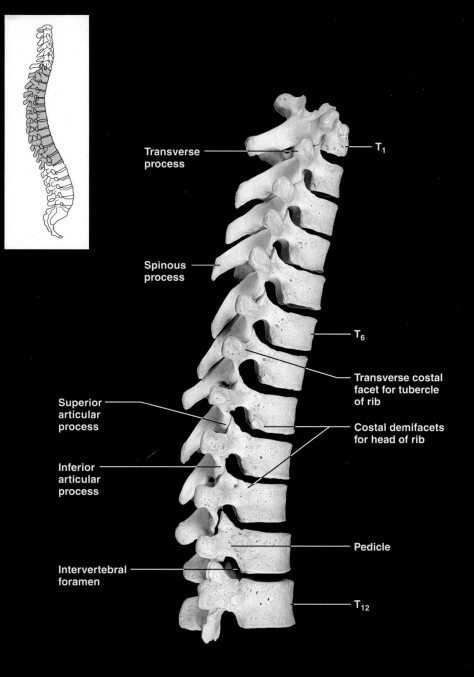

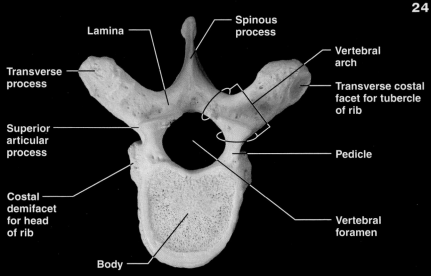

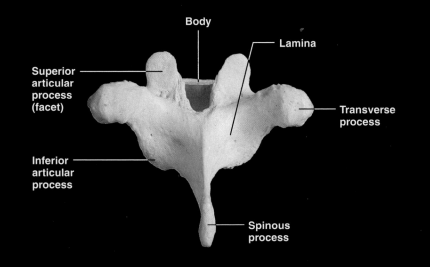

**(a) articulated thoracic vertebrae,
right lateral view**

**(b) seventh (typical) thoracic vertebra,
superior view**

**(c) seventh (typical) thoracic vertebra,
posterior view**

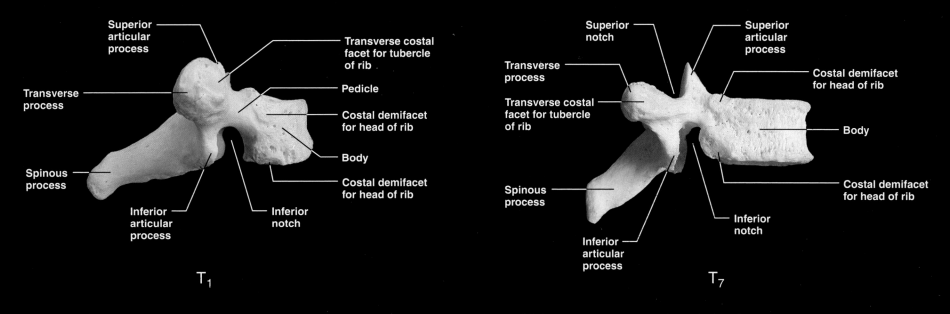

**T₁**

Superior articular process

Transverse costal facet for tubercle of rib

Pedicle

Costal demifacet for head of rib

Body

Costal demifacet for head of rib

Transverse process

Spinous process

Inferior articular process

Inferior notch

**T₇**

Superior notch

Superior articular process

Transverse process

Costal demifacet for head of rib

Transverse costal facet for tubercle of rib

Body

Spinous process

Costal demifacet for head of rib

Inferior articular process

Inferior notch

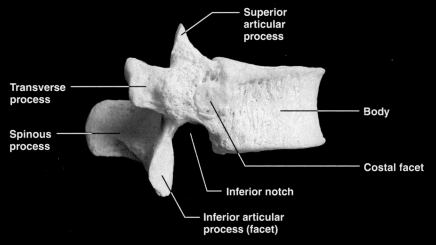

**T₁₂**

Superior articular process

Transverse process

Body

Spinous process

Costal facet

Inferior notch

Inferior articular process (facet)

(d) comparison of T₁, T₇, and T₁₂ in right lateral views

**Figure 20** Thoracic vertebrae.

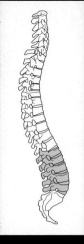

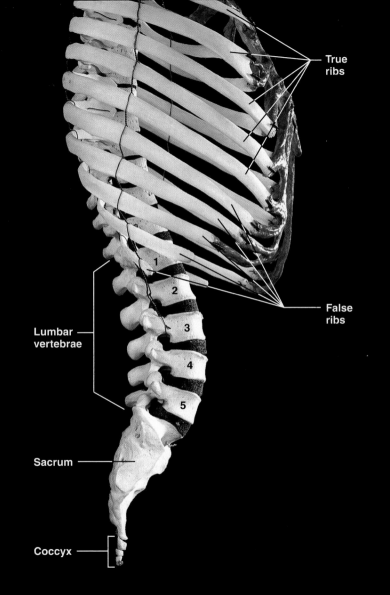

True ribs

False ribs

Lumbar vertebrae

1
2
3
4
5

Sacrum

Coccyx

(a) articulated lumbar vertebrae and rib cage,

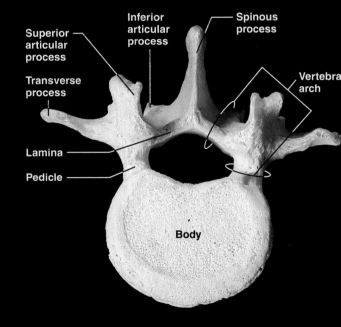

Superior articular process

Transverse process

Lamina

Pedicle

Inferior articular process

Spinous process

Vertebral arch

Body

(b) second lumbar vertebra,

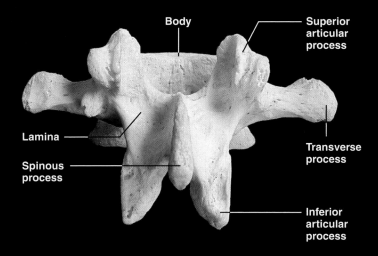

Body

Superior
articular
process

Lamina

Transverse
process

Spinous
process

Inferior
articular
process

(c) second lumbar vertebra, posterior view

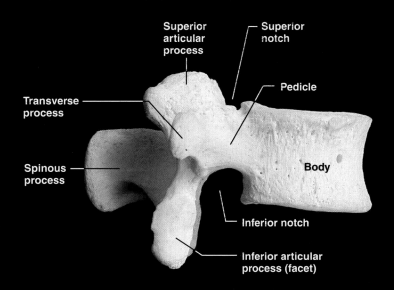

Superior
articular
process

Superior
notch

Pedicle

Transverse
process

Spinous
process

Body

Inferior notch

Inferior articular
process (facet)

(d) second lumbar vertebra, right lateral view

**Figure 21**   Lumbar vertebrae.

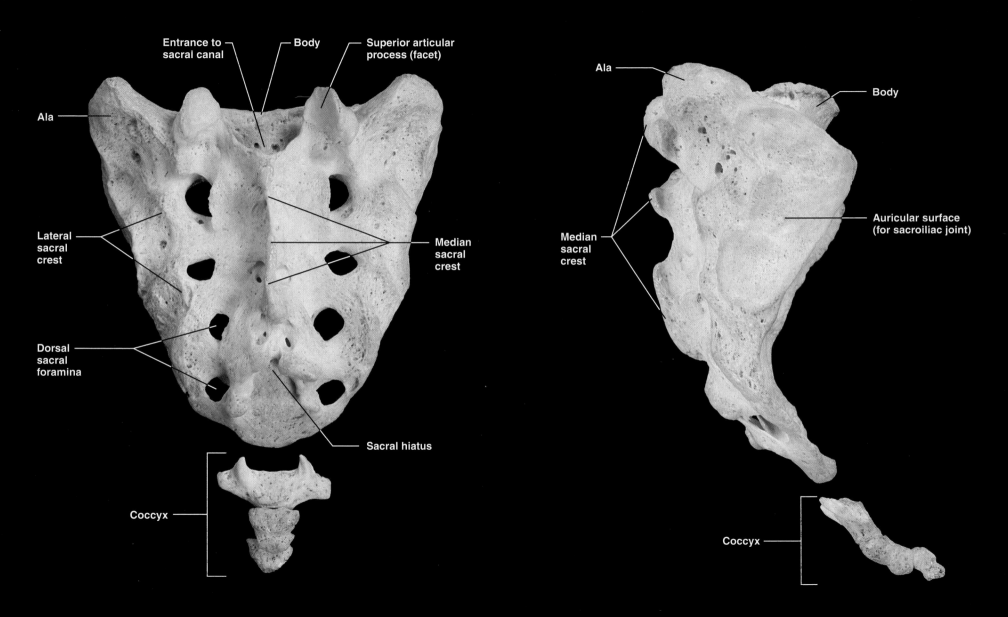

Entrance to sacral canal

Body

Superior articular process (facet)

Ala

Lateral sacral crest

Dorsal sacral foramina

Median sacral crest

Sacral hiatus

Coccyx

(a) posterior view

Ala

Body

Median sacral crest

Auricular surface (for sacroiliac joint)

Coccyx

(b) right lateral view

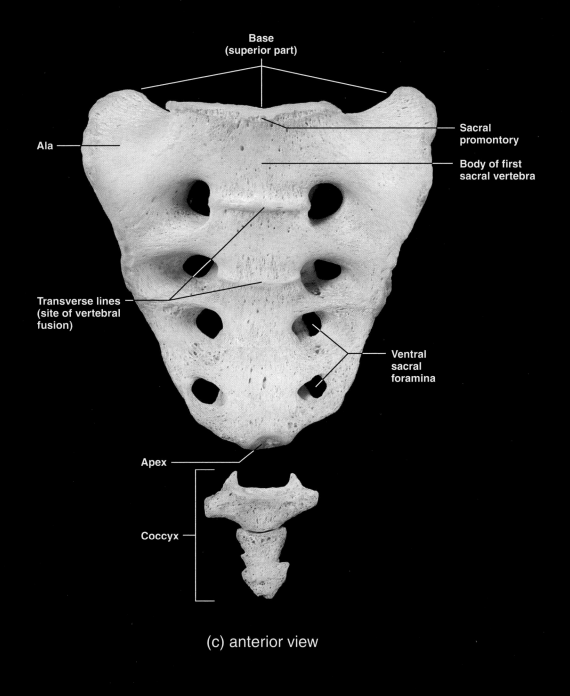

Base
(superior part)

Ala

Sacral
promontory

Body of first
sacral vertebra

Transverse lines
(site of vertebral
fusion)

Ventral
sacral
foramina

Apex

Coccyx

(c) anterior view

**Figure 22** Sacrum and coccyx.

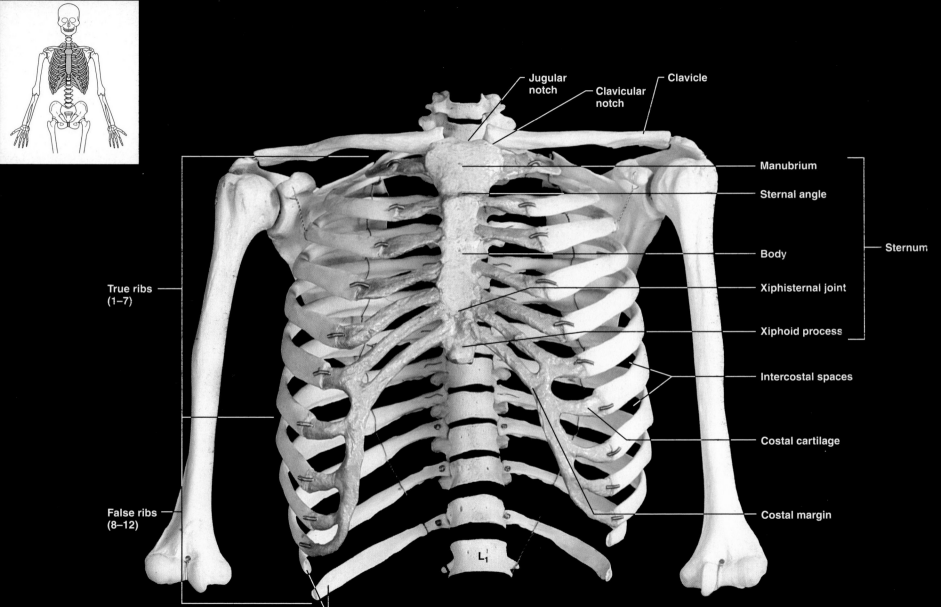

Jugular notch

Clavicular notch

Clavicle

Manubrium

Sternal angle

Body

Xiphisternal joint

Xiphoid process

Sternum

Intercostal spaces

Costal cartilage

Costal margin

True ribs (1–7)

False ribs (8–12)

L₁

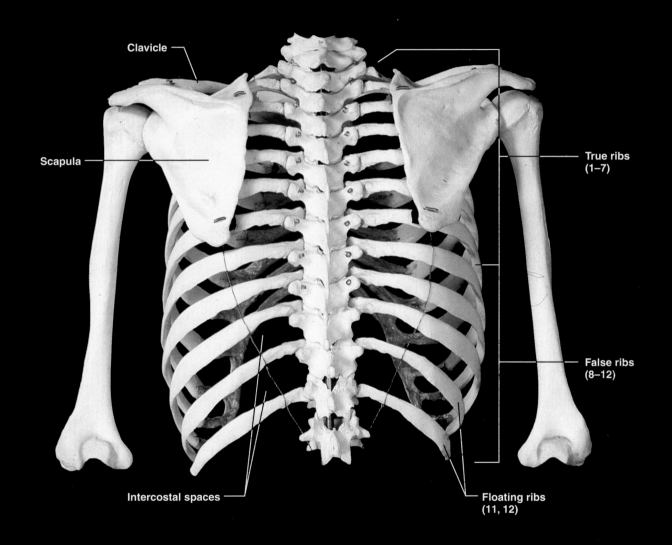

Clavicle

Scapula

True ribs
(1–7)

False ribs
(8–12)

Intercostal spaces

Floating ribs
(11, 12)

(b) posterior view

**Figure 23**   Bony thorax.

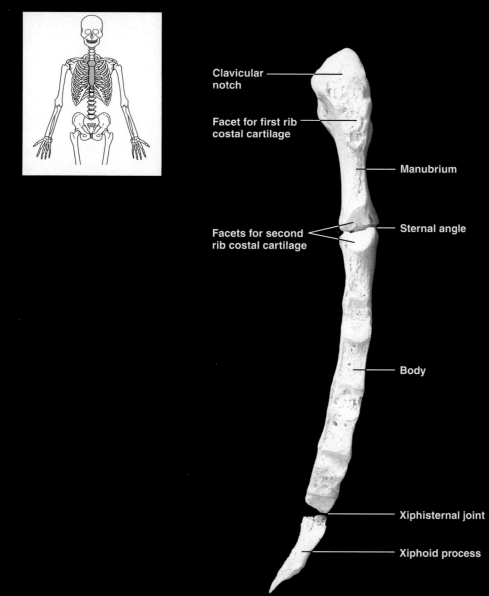

Clavicular notch

Facet for first rib costal cartilage

Manubrium

Facets for second rib costal cartilage

Sternal angle

Body

Xiphisternal joint

Xiphoid process

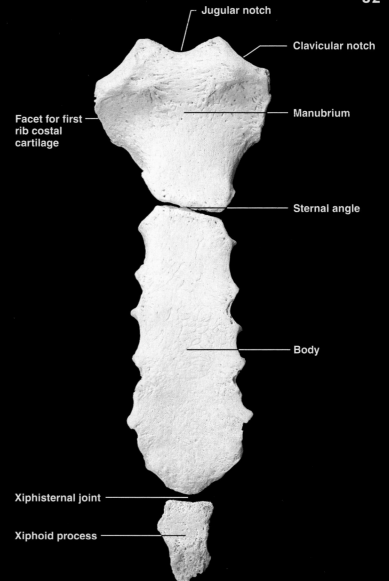

Jugular notch

Clavicular notch

Facet for first rib costal cartilage

Manubrium

Sternal angle

Body

Xiphisternal joint

Xiphoid process

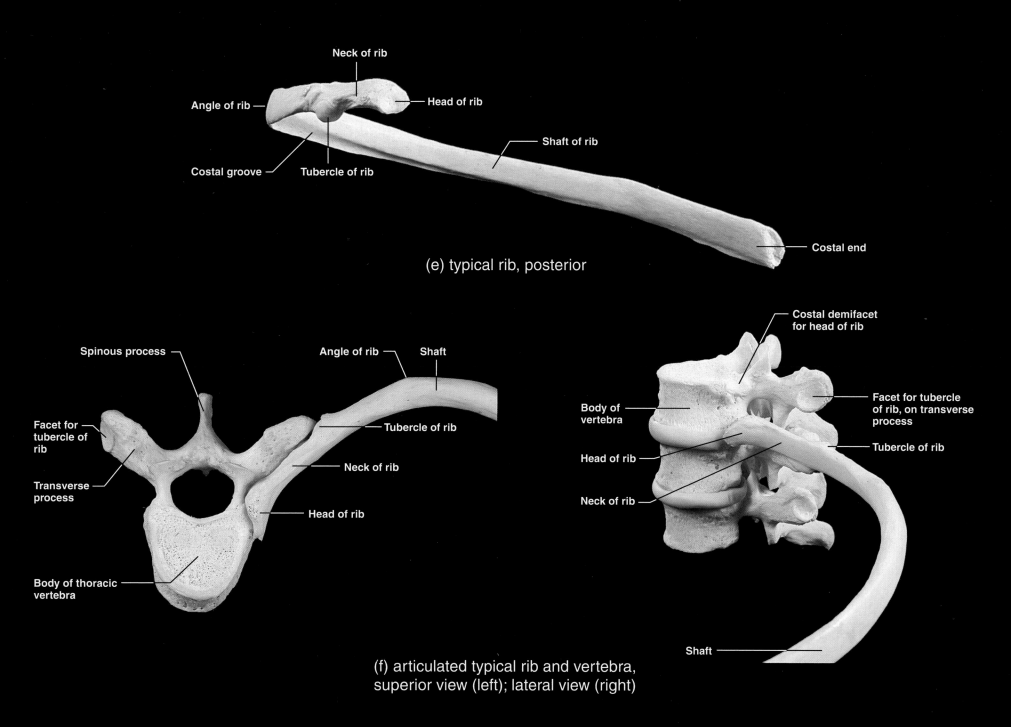

(e) typical rib, posterior

(f) articulated typical rib and vertebra,
superior view (left); lateral view (right)

**Figure 23** Bony thorax (continued).

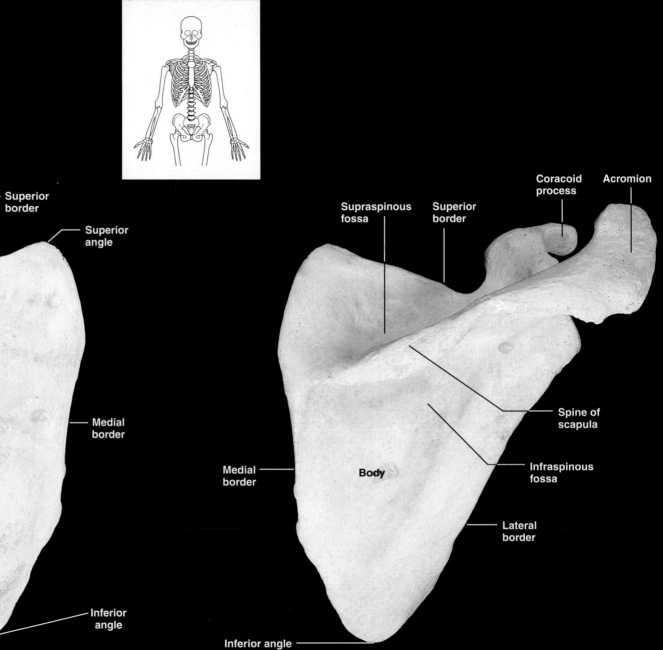

**Acromion** **Coracoid process** **Scapular notch** **Superior border**

**Superior angle**

**Glenoid cavity**

**Lateral angle**

**Medial border**

**Subscapular fossa**

**Lateral border**

**Inferior angle**

**(a) right scapula, anterior view**

**Supraspinous fossa** **Superior border** **Coracoid process** **Acromion**

**Medial border**

**Body**

**Spine of scapula**

**Infraspinous fossa**

**Lateral border**

**Inferior angle**

**(b) right scapula, posterior view**

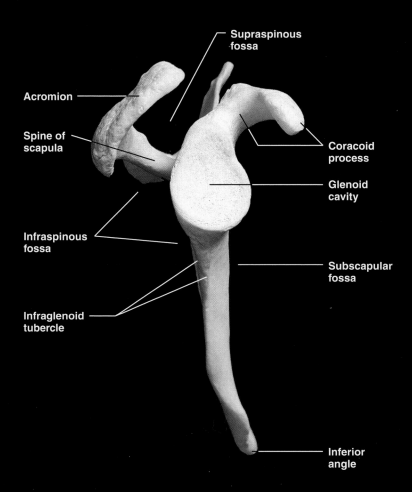

Supraspinous fossa

Acromion

Spine of scapula

Coracoid process

Glenoid cavity

Infraspinous fossa

Subscapular fossa

Infraglenoid tubercle

Inferior angle

(c) right scapula, lateral aspect

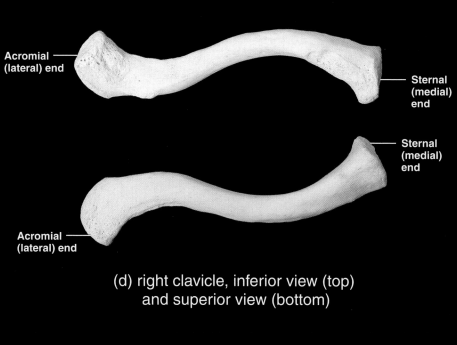

Acromial (lateral) end

Sternal (medial) end

Sternal (medial) end

Acromial (lateral) end

(d) right clavicle, inferior view (top) and superior view (bottom)

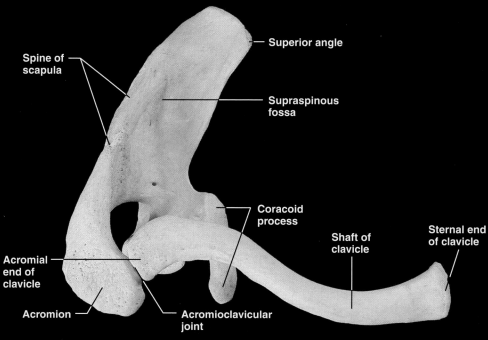

Spine of scapula

Superior angle

Supraspinous fossa

Coracoid process

Shaft of clavicle

Sternal end of clavicle

Acromial end of clavicle

Acromion

Acromioclavicular joint

(e) articulated right clavicle and scapula, superior view

**Figure 24** Scapula and clavicle.

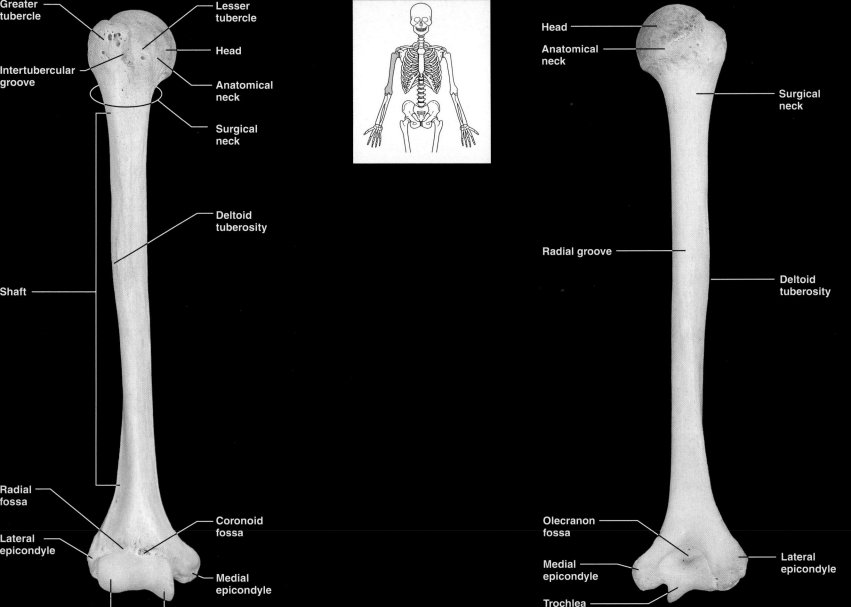

Greater tubercle

Lesser tubercle

Head

Intertubercular groove

Anatomical neck

Surgical neck

Deltoid tuberosity

Shaft

Radial fossa

Lateral epicondyle

Coronoid fossa

Medial epicondyle

Head

Anatomical neck

Surgical neck

Radial groove

Deltoid tuberosity

Olecranon fossa

Medial epicondyle

Lateral epicondyle

Trochlea

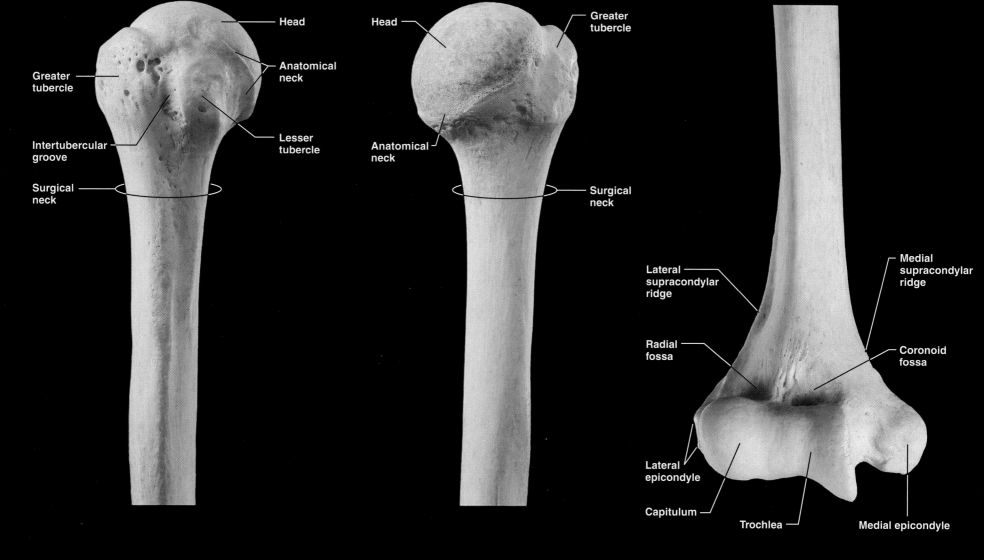

Head

Anatomical neck

Greater tubercle

Lesser tubercle

Intertubercular groove

Surgical neck

Head

Greater tubercle

Anatomical neck

Surgical neck

Lateral supracondylar ridge

Medial supracondylar ridge

Radial fossa

Coronoid fossa

Lateral epicondyle

Capitulum

Trochlea

Medial epicondyle

(c) proximal end, anterior view

(d) proximal end, posterior view

(e) distal end, anterior view

**Figure 25** Right humerus.

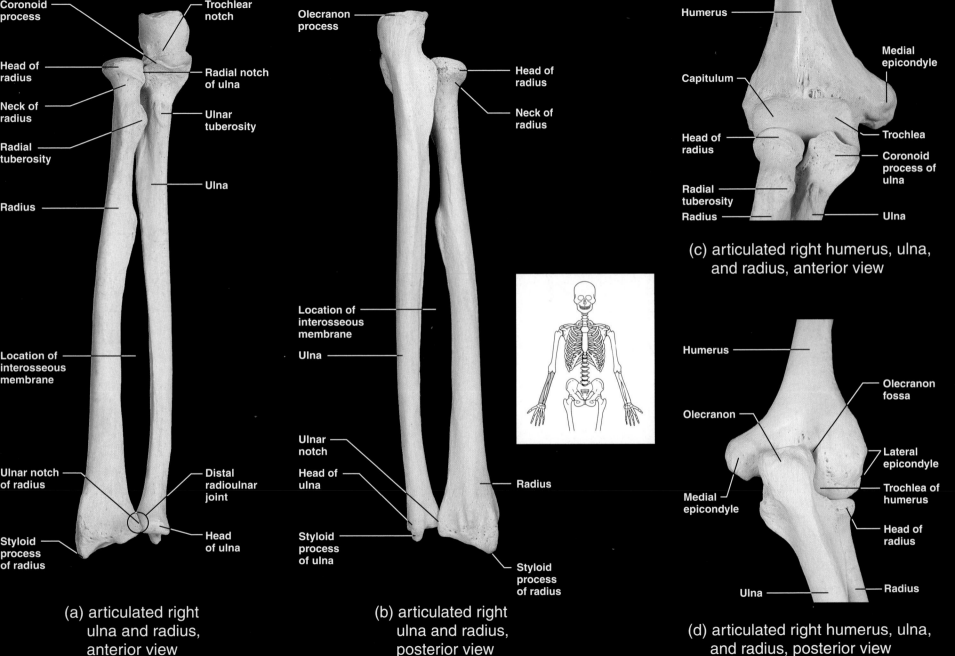

Coronoid process

Trochlear notch

Head of radius

Radial notch of ulna

Neck of radius

Ulnar tuberosity

Radial tuberosity

Ulna

Radius

Location of interosseous membrane

Ulnar notch of radius

Distal radioulnar joint

Styloid process of radius

Head of ulna

(a) articulated right ulna and radius, anterior view

Olecranon process

Head of radius

Neck of radius

Location of interosseous membrane

Ulna

Ulnar notch

Head of ulna

Radius

Styloid process of ulna

Styloid process of radius

(b) articulated right ulna and radius, posterior view

Humerus

Medial epicondyle

Capitulum

Head of radius

Trochlea

Coronoid process of ulna

Radial tuberosity

Radius

Ulna

(c) articulated right humerus, ulna, and radius, anterior view

Humerus

Olecranon fossa

Olecranon

Lateral epicondyle

Medial epicondyle

Trochlea of humerus

Head of radius

Ulna

Radius

(d) articulated right humerus, ulna, and radius, posterior view

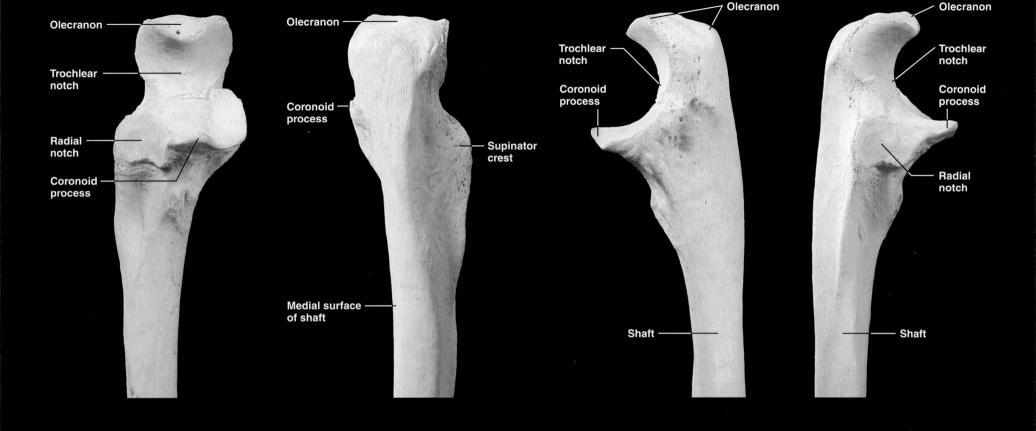

(e) right ulna, proximal end

**Figure 26** Right ulna and radius.

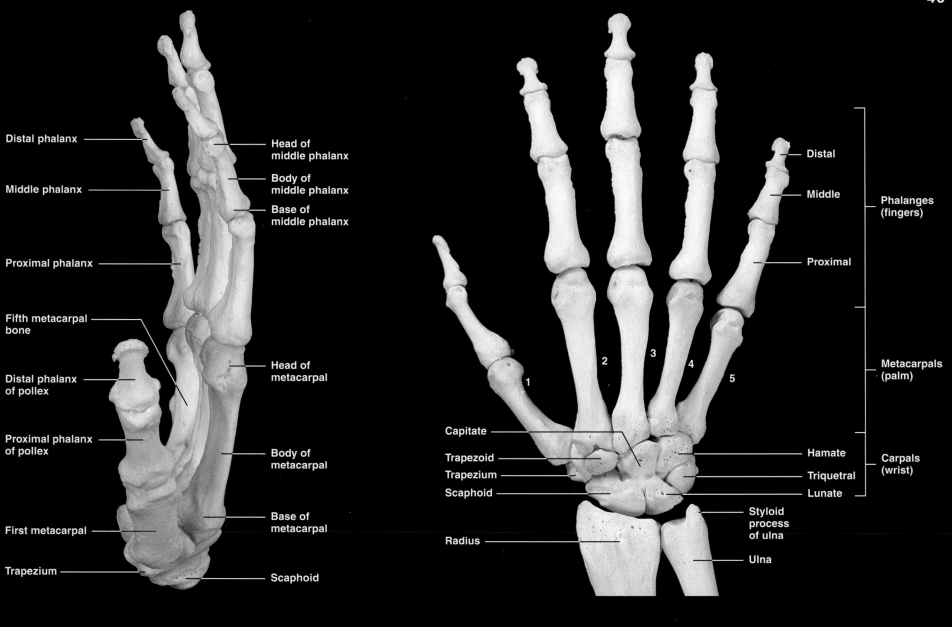

Distal phalanx

Middle phalanx

Proximal phalanx

Fifth metacarpal
bone

Distal phalanx
of pollex

Proximal phalanx
of pollex

First metacarpal

Trapezium

Head of
middle phalanx

Body of
middle phalanx

Base of
middle phalanx

Head of
metacarpal

Body of
metacarpal

Base of
metacarpal

Scaphoid

(a) lateral aspect

Distal

Middle

Proximal

Phalanges
(fingers)

Metacarpals
(palm)

Capitate

Trapezoid

Trapezium

Scaphoid

Radius

Hamate

Triquetral

Lunate

Styloid
process
of ulna

Ulna

Carpals
(wrist)

1   2   3   4   5

(b) dorsal aspect

**Figure 27**   Bones of the right hand.

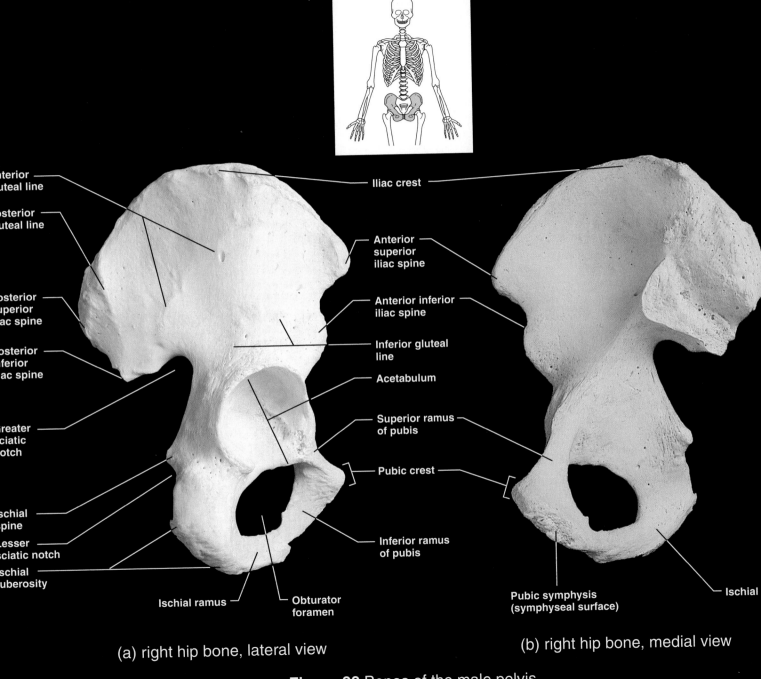

Anterior
gluteal line

Posterior
gluteal line

Posterior
superior
iliac spine

Posterior
inferior
iliac spine

Greater
sciatic
notch

Ischial
spine

Lesser
sciatic notch

Ischial
tuberosity

Iliac crest

Anterior
superior
iliac spine

Anterior inferior
iliac spine

Inferior gluteal
line

Acetabulum

Superior ramus
of pubis

Pubic crest

Inferior ramus
of pubis

Ischial ramus

Obturator
foramen

Pubic symphysis
(symphyseal surface)

Ischial

(a) right hip bone, lateral view

(b) right hip bone, medial view

Figure 26 Bones of the male pelvis

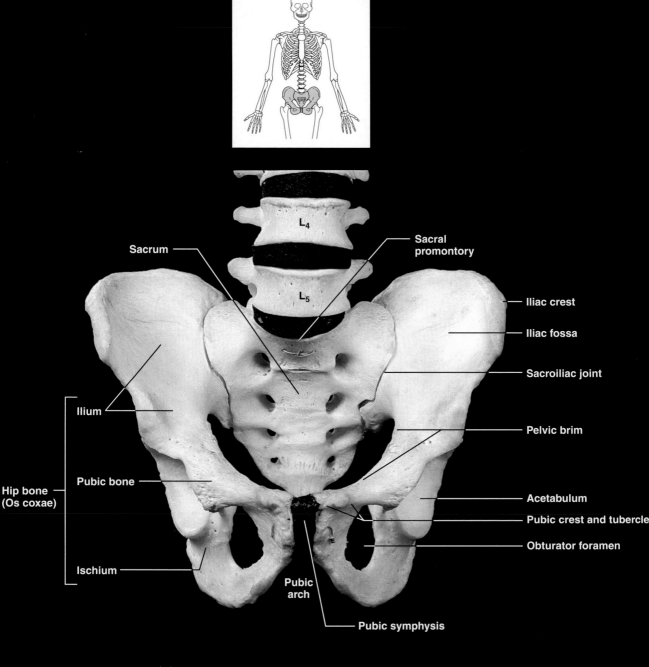

L₄

L₅

Sacrum

Sacral promontory

Iliac crest

Iliac fossa

Sacroiliac joint

Pelvic brim

Ilium

Hip bone (Os coxae)

Pubic bone

Acetabulum

Pubic crest and tubercle

Obturator foramen

Ischium

Pubic arch

Pubic symphysis

(c) articulated male pelvis, anterior view

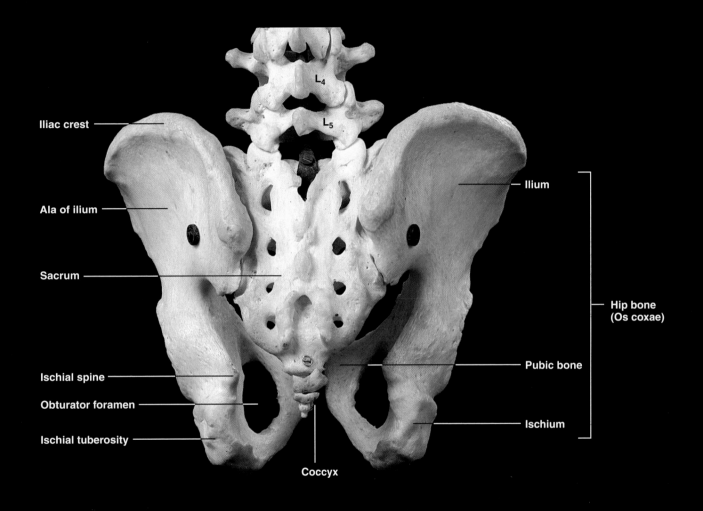

Iliac crest

Ala of ilium

Sacrum

Ischial spine

Obturator foramen

Ischial tuberosity

L₄

L₅

Ilium

Hip bone
(Os coxae)

Pubic bone

Ischium

Coccyx

(d) articulated male pelvis, posterior view

**Figure 28** Bones of the male pelvis (continued).

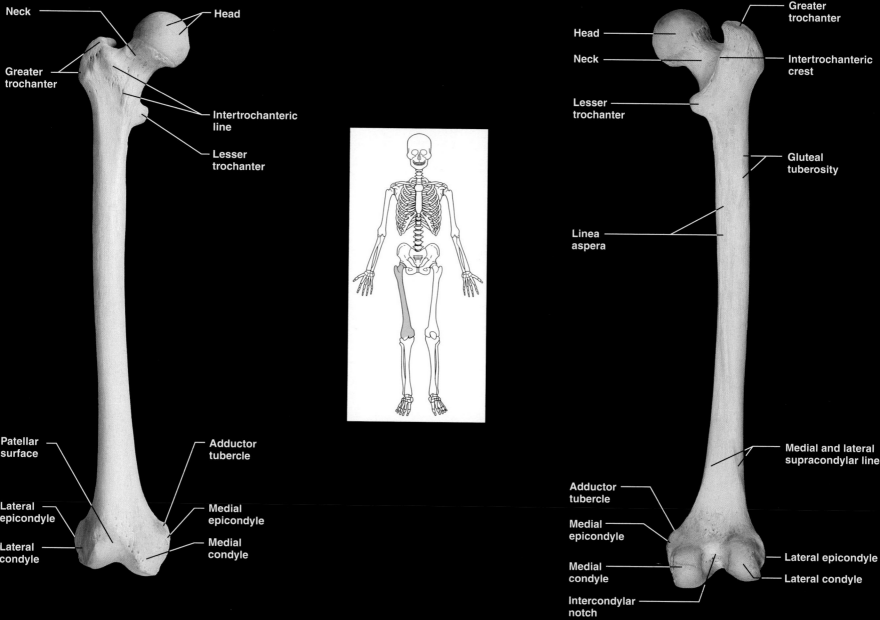

Neck

Head

Greater
trochanter

Intertrochanteric
line

Lesser
trochanter

Patellar
surface

Adductor
tubercle

Lateral
epicondyle

Medial
epicondyle

Lateral
condyle

Medial
condyle

Head

Greater
trochanter

Neck

Intertrochanteric
crest

Lesser
trochanter

Gluteal
tuberosity

Linea
aspera

Medial and lateral
supracondylar line

Adductor
tubercle

Medial
epicondyle

Lateral epicondyle

Medial
condyle

Lateral condyle

Intercondylar
notch

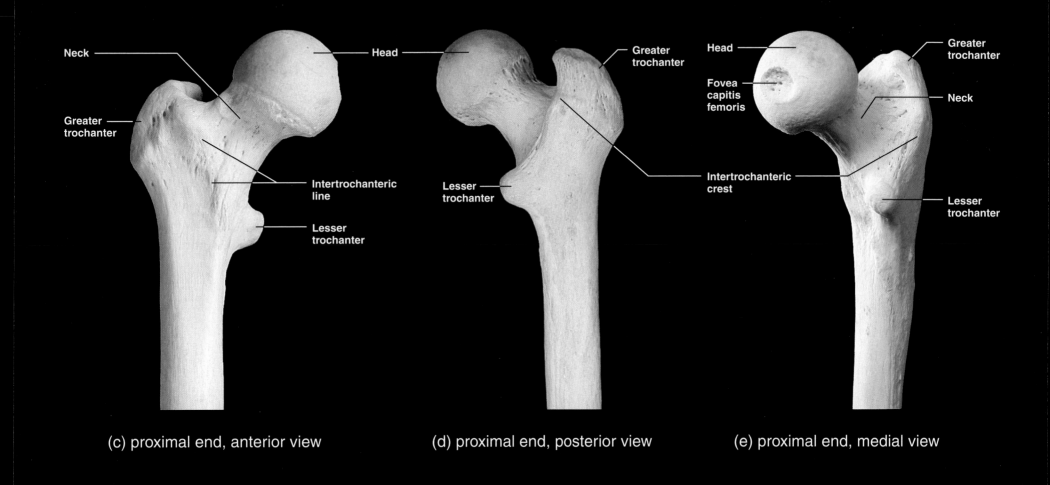

Neck

Head

Greater
trochanter

Greater
trochanter

Intertrochanteric
line

Lesser
trochanter

Lesser
trochanter

Greater
trochanter

Head

Fovea
capitis
femoris

Neck

Intertrochanteric
crest

Lesser
trochanter

(c) proximal end, anterior view

(d) proximal end, posterior view

(e) proximal end, medial view

**Figure 29** Right femur.

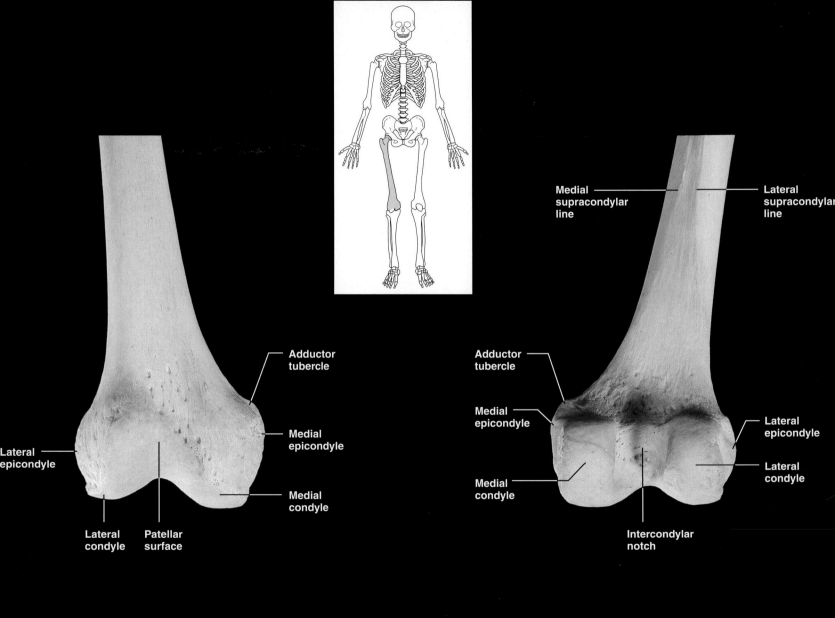

**Adductor tubercle**

**Medial epicondyle**

**Medial condyle**

**Lateral epicondyle**

**Lateral condyle**

**Patellar surface**

(f) distal end, anterior view

**Medial supracondylar line**

**Lateral supracondylar line**

**Adductor tubercle**

**Medial epicondyle**

**Medial condyle**

**Lateral epicondyle**

**Lateral condyle**

**Intercondylar notch**

(g) distal end, posterior view

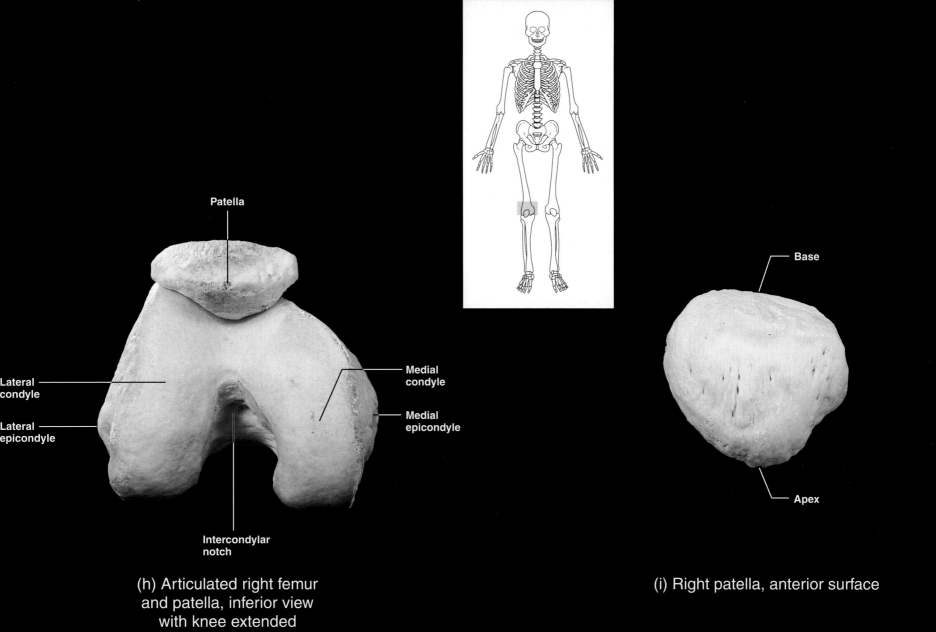

Patella

Lateral
condyle

Lateral
epicondyle

Medial
condyle

Medial
epicondyle

Intercondylar
notch

(h) Articulated right femur
and patella, inferior view
with knee extended

Base

Apex

(i) Right patella, anterior surface

**Figure 29** Right femur (continued).

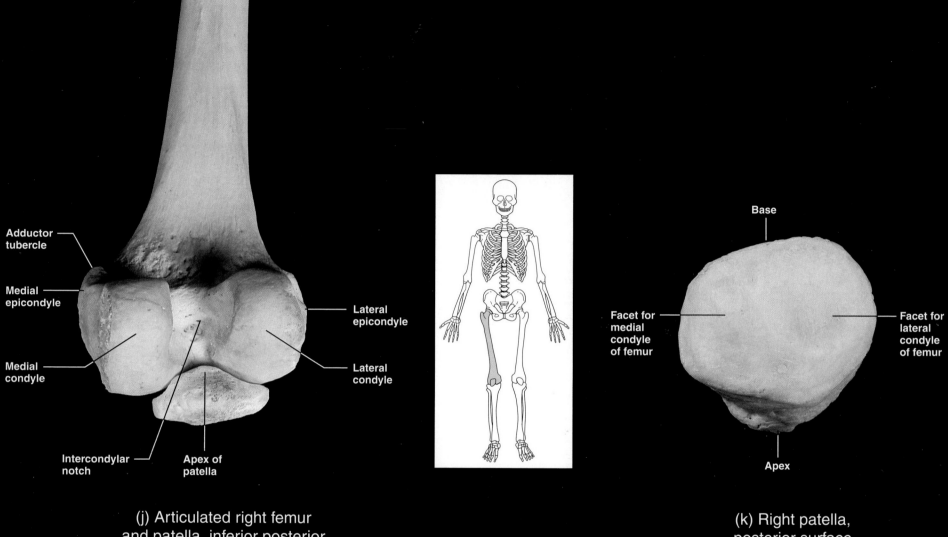

**Adductor tubercle**

**Medial epicondyle**

**Medial condyle**

**Intercondylar notch**

**Apex of patella**

**Lateral epicondyle**

**Lateral condyle**

(j) Articulated right femur
and patella, inferior posterior
view with knee flexed

**Base**

**Facet for medial condyle of femur**

**Facet for lateral condyle of femur**

**Apex**

(k) Right patella,
posterior surface

**Figure 29**   Right femur (continued).

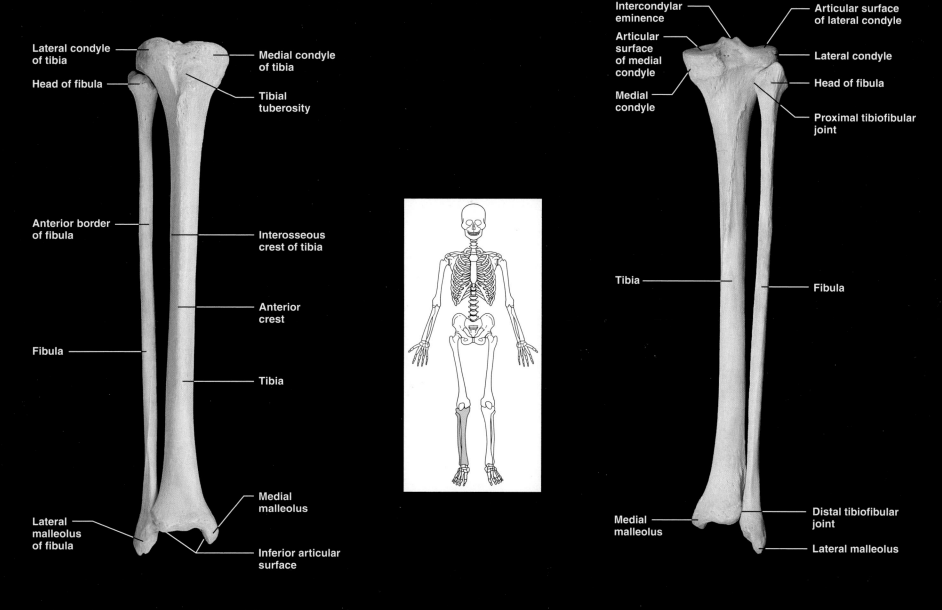

**Lateral condyle of tibia**

**Head of fibula**

**Anterior border of fibula**

**Fibula**

**Lateral malleolus of fibula**

**Medial condyle of tibia**

**Tibial tuberosity**

**Interosseous crest of tibia**

**Anterior crest**

**Tibia**

**Medial malleolus**

**Inferior articular surface**

(a) articulated right tibia and fibula,
anterior view

**Intercondylar eminence**

**Articular surface of medial condyle**

**Medial condyle**

**Articular surface of lateral condyle**

**Lateral condyle**

**Head of fibula**

**Proximal tibiofibular joint**

**Tibia**

**Fibula**

**Medial malleolus**

**Distal tibiofibular joint**

**Lateral malleolus**

(b) articulated right tibia and fibula,
posterior view

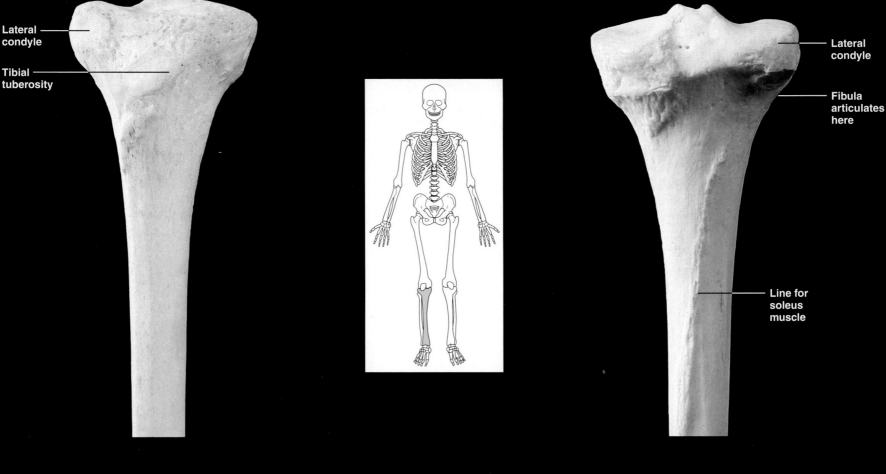

Lateral condyle

Tibial tuberosity

Lateral condyle

Fibula articulates here

Line for soleus muscle

(c) right tibia, proximal end, anterior view

(d) right tibia, proximal end, posterior view

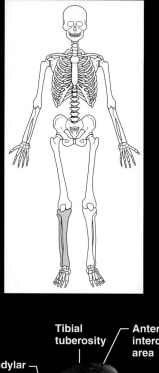

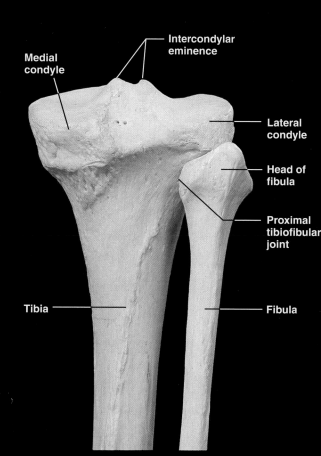

Medial
condyle

Intercondylar
eminence

Lateral
condyle

Head of
fibula

Proximal
tibiofibular
joint

Tibia

Fibula

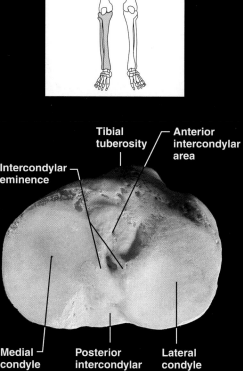

Intercondylar
eminence

Tibial
tuberosity

Anterior
intercondylar
area

Medial
condyle

Posterior
intercondylar
area

Lateral
condyle

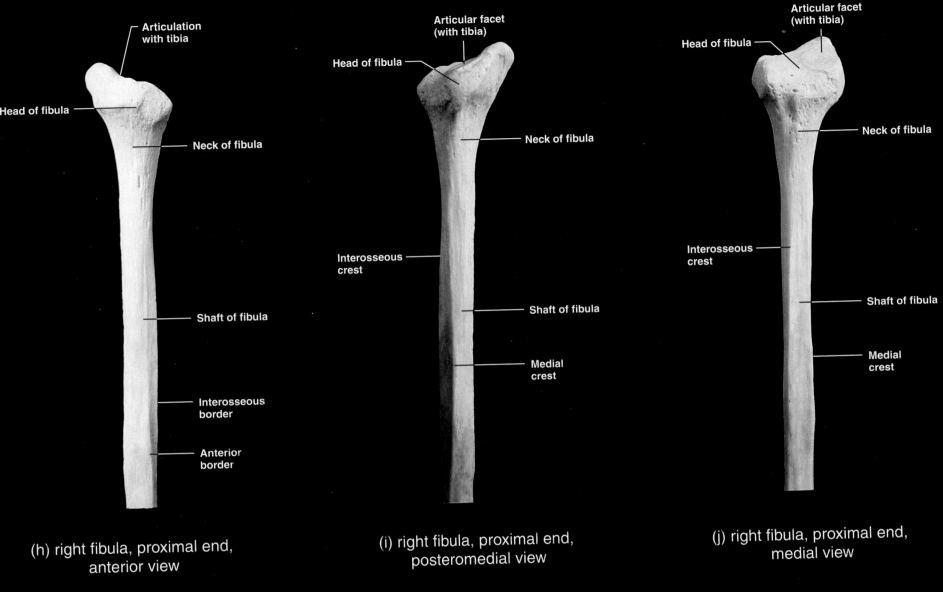

Head of fibula

Articulation with tibia

Neck of fibula

Shaft of fibula

Interosseous border

Anterior border

(h) right fibula, proximal end, anterior view

Articular facet (with tibia)

Head of fibula

Neck of fibula

Interosseous crest

Shaft of fibula

Medial crest

(i) right fibula, proximal end, posteromedial view

Articular facet (with tibia)

Head of fibula

Neck of fibula

Interosseous crest

Shaft of fibula

Medial crest

(j) right fibula, proximal end, medial view

**Figure 30**  Right tibia and fibula (continued).

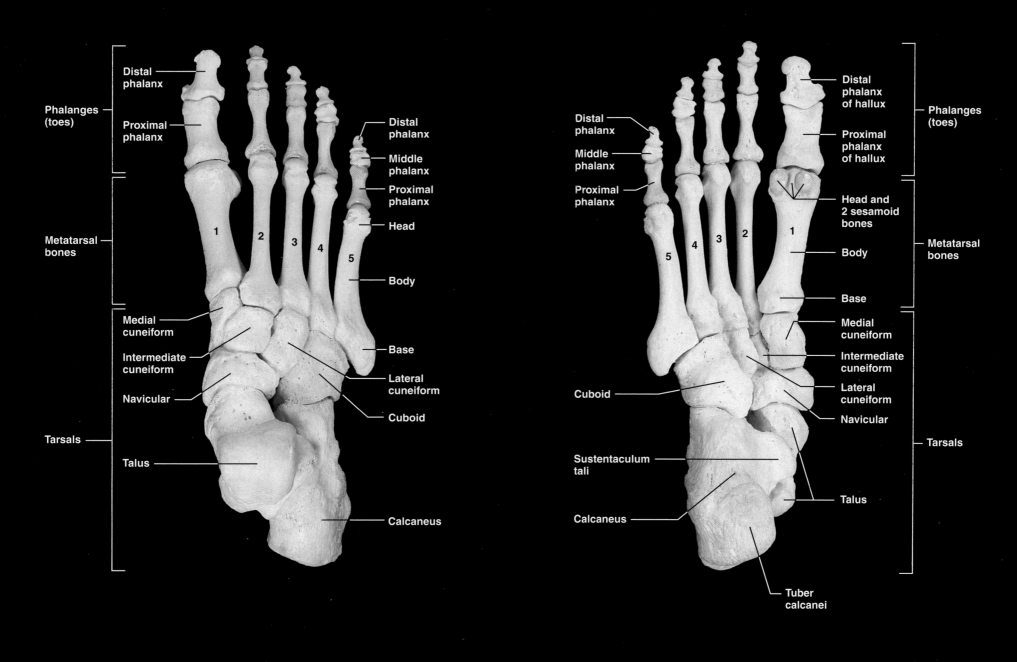

**Figure 31** Bones of the right ankle and foot.

(a) superior surface

(b) inferior (plantar) surface

53

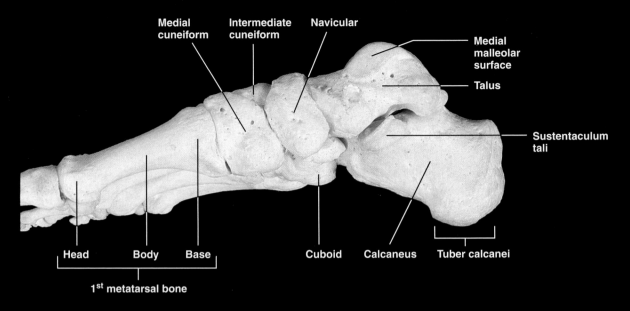

Medial cuneiform
Intermediate cuneiform
Navicular
Medial malleolar surface
Talus
Sustentaculum tali
Head
Body
Base
Cuboid
Calcaneus
Tuber calcanei
1st metatarsal bone

(c) medial view

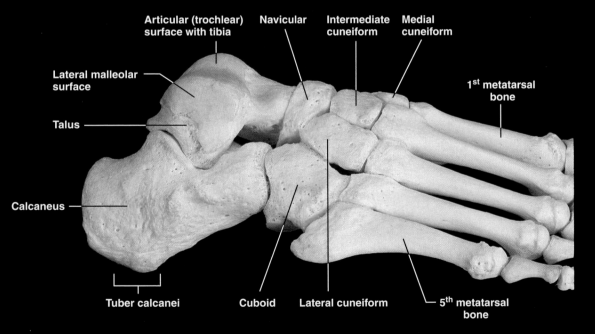

Articular (trochlear) surface with tibia
Navicular
Intermediate cuneiform
Medial cuneiform
Lateral malleolar surface
Talus
1st metatarsal bone
Calcaneus
Tuber calcanei
Cuboid
Lateral cuneiform
5th metatarsal bone

(d) lateral view

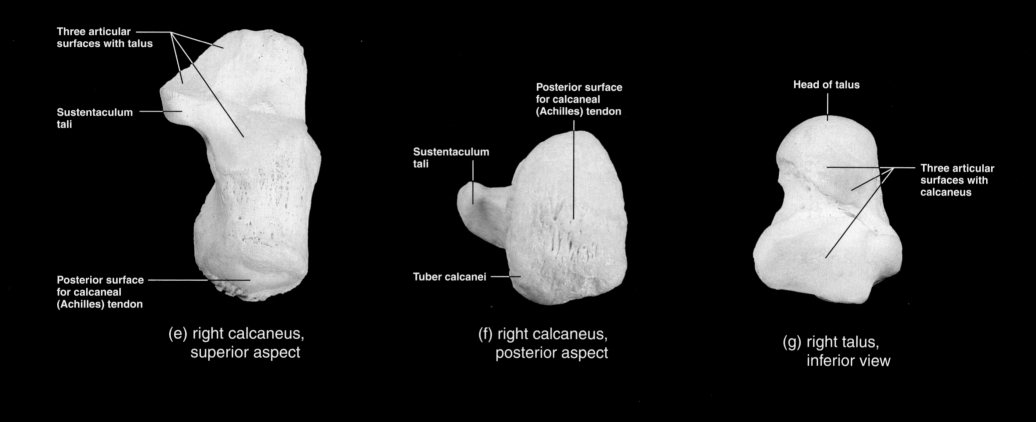

**Three articular surfaces with talus**

**Sustentaculum tali**

**Posterior surface for calcaneal (Achilles) tendon**

(e) right calcaneus, superior aspect

**Posterior surface for calcaneal (Achilles) tendon**

**Sustentaculum tali**

**Tuber calcanei**

(f) right calcaneus, posterior aspect

**Head of talus**

**Three articular surfaces with calcaneus**

(g) right talus, inferior view

**Figure 31** Bones of the right ankle and foot (continued).

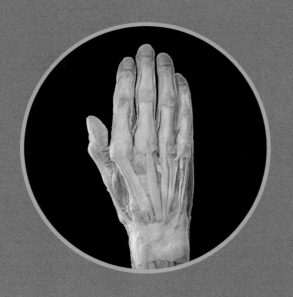

# Part II

# Soft Tissue of the Human Body

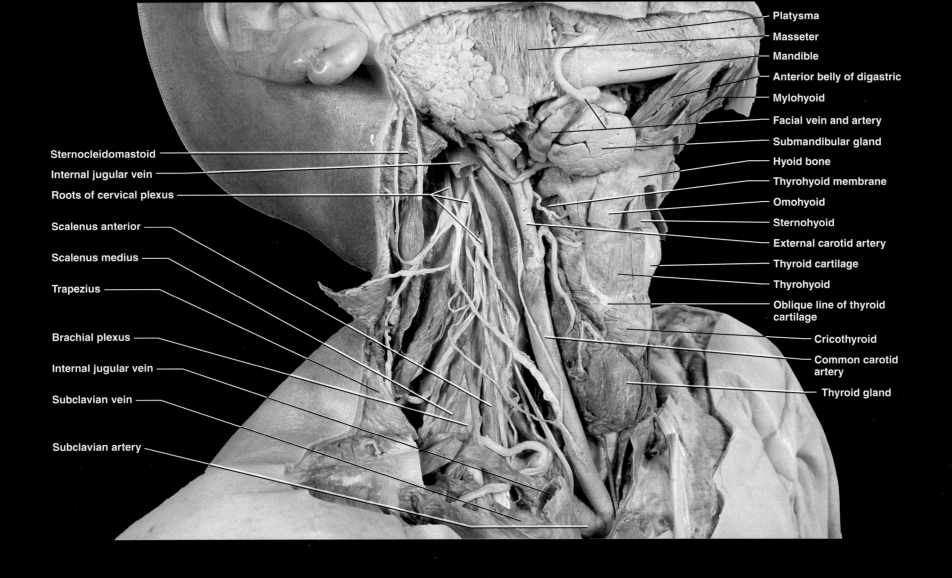

**Figure 32** Right lower face and upper neck.

Platysma
Masseter
Mandible
Anterior belly of digastric
Mylohyoid
Facial vein and artery
Submandibular gland
Hyoid bone
Thyrohyoid membrane
Omohyoid
Sternohyoid
External carotid artery
Thyroid cartilage
Thyrohyoid
Oblique line of thyroid cartilage
Cricothyroid
Common carotid artery
Thyroid gland

Sternocleidomastoid
Internal jugular vein
Roots of cervical plexus
Scalenus anterior
Scalenus medius
Trapezius
Brachial plexus
Internal jugular vein
Subclavian vein
Subclavian artery

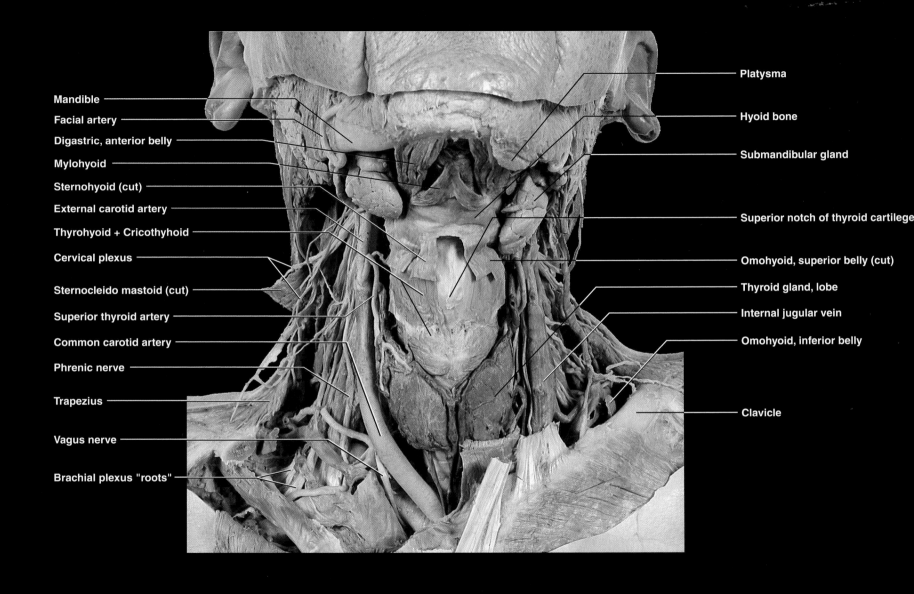

Mandible

Facial artery

Digastric, anterior belly

Mylohyoid

Sternohyoid (cut)

External carotid artery

Thyrohyoid + Cricothyhoid

Cervical plexus

Sternocleido mastoid (cut)

Superior thyroid artery

Common carotid artery

Phrenic nerve

Trapezius

Vagus nerve

Brachial plexus "roots"

Platysma

Hyoid bone

Submandibular gland

Superior notch of thyroid cartilege

Omohyoid, superior belly (cut)

Thyroid gland, lobe

Internal jugular vein

Omohyoid, inferior belly

Clavicle

**Figure 33** Muscles, blood vessels, and nerves of neck, anterior view.

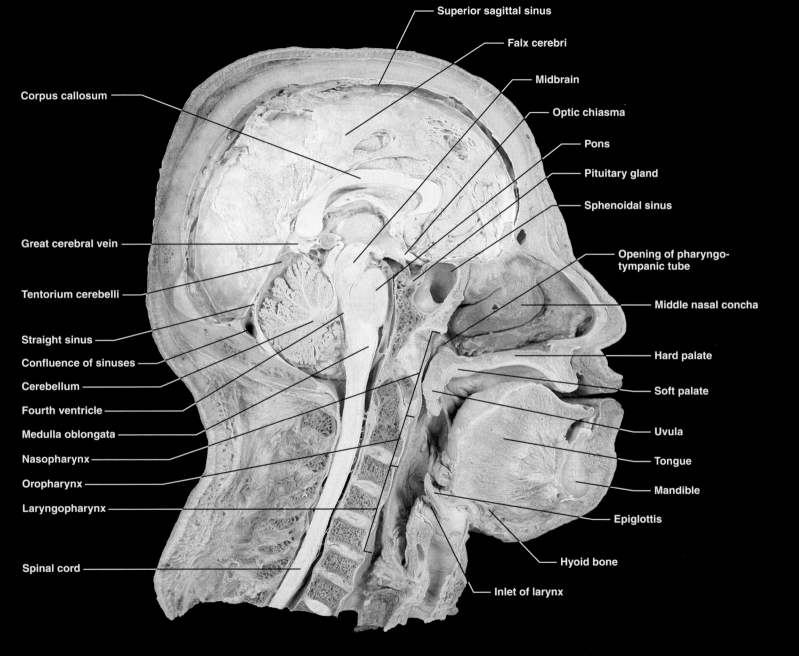

Superior sagittal sinus

Falx cerebri

Midbrain

Corpus callosum

Optic chiasma

Pons

Pituitary gland

Sphenoidal sinus

Great cerebral vein

Opening of pharyngo-tympanic tube

Tentorium cerebelli

Middle nasal concha

Straight sinus

Hard palate

Confluence of sinuses

Cerebellum

Soft palate

Fourth ventricle

Medulla oblongata

Uvula

Nasopharynx

Tongue

Oropharynx

Mandible

Laryngopharynx

Epiglottis

Hyoid bone

Spinal cord

Inlet of larynx

**Figure 34** Sagittal section of the head.

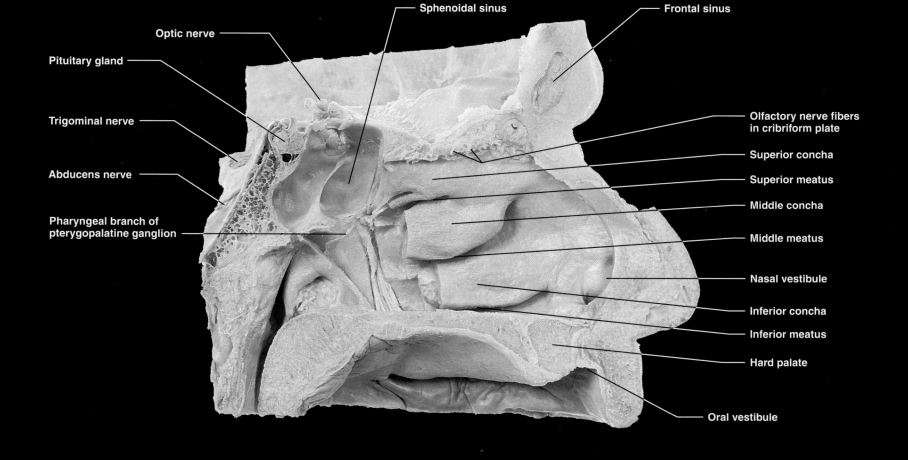

**Figure 35**  Left nasal cavity, lateral wall.

Sphenoidal sinus

Frontal sinus

Optic nerve

Pituitary gland

Trigominal nerve

Abducens nerve

Pharyngeal branch of
pterygopalatine ganglion

Olfactory nerve fibers
in cribriform plate

Superior concha

Superior meatus

Middle concha

Middle meatus

Nasal vestibule

Inferior concha

Inferior meatus

Hard palate

Oral vestibule

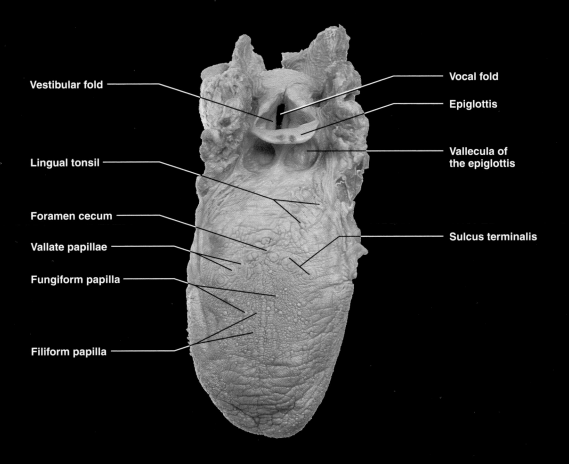

Vestibular fold

Vocal fold

Epiglottis

Lingual tonsil

Vallecula of
the epiglottis

Foramen cecum

Vallate papillae

Sulcus terminalis

Fungiform papilla

Filiform papilla

**Figure 36**   Tongue and laryngeal inlet.

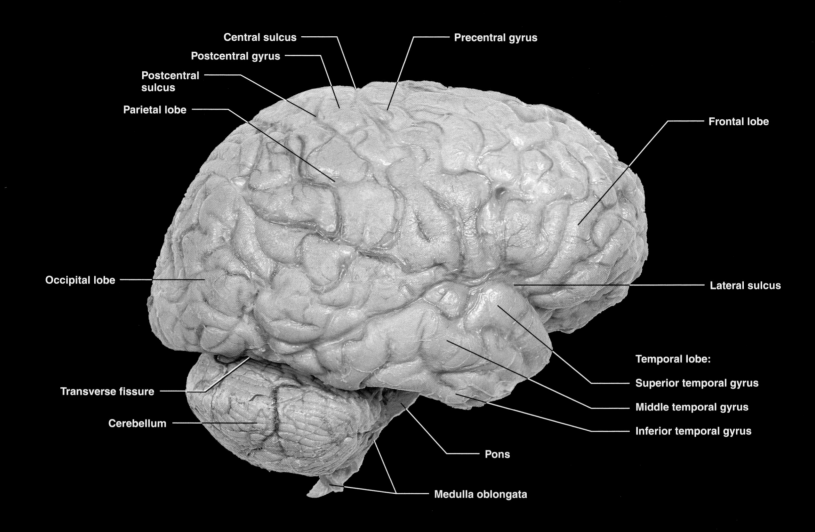

Central sulcus

Precentral gyrus

Postcentral gyrus

Postcentral sulcus

Parietal lobe

Frontal lobe

Occipital lobe

Lateral sulcus

Temporal lobe:

Superior temporal gyrus

Transverse fissure

Middle temporal gyrus

Cerebellum

Inferior temporal gyrus

Pons

Medulla oblongata

**Figure 37**   Right cerebral hemisphere (arachnoid mater removed).

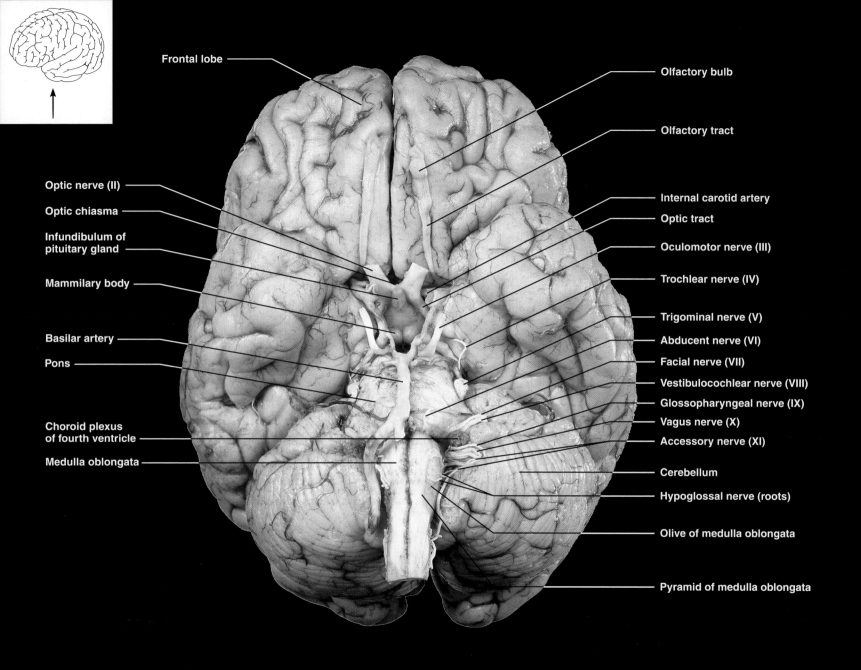

Frontal lobe

Olfactory bulb

Olfactory tract

Optic nerve (II)

Internal carotid artery

Optic chiasma

Optic tract

Infundibulum of
pituitary gland

Oculomotor nerve (III)

Trochlear nerve (IV)

Mammilary body

Trigominal nerve (V)

Basilar artery

Abducent nerve (VI)

Pons

Facial nerve (VII)

Vestibulocochlear nerve (VIII)

Glossopharyngeal nerve (IX)

Vagus nerve (X)

Choroid plexus
of fourth ventricle

Accessory nerve (XI)

Medulla oblongata

Cerebellum

Hypoglossal nerve (roots)

Olive of medulla oblongata

Pyramid of medulla oblongata

**Figure 38** Ventral view of the brain.

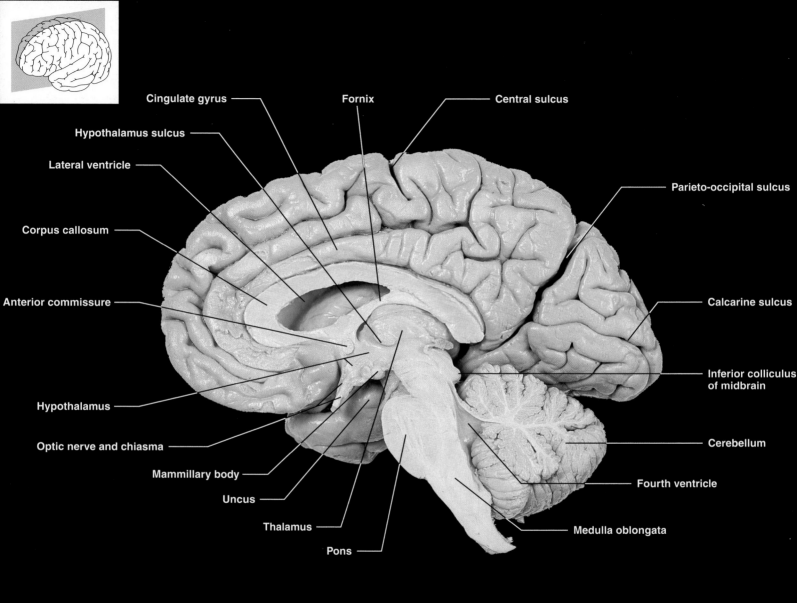

Cingulate gyrus

Fornix

Central sulcus

Hypothalamus sulcus

Lateral ventricle

Parieto-occipital sulcus

Corpus callosum

Anterior commissure

Calcarine sulcus

Inferior colliculus of midbrain

Hypothalamus

Optic nerve and chiasma

Cerebellum

Mammillary body

Fourth ventricle

Uncus

Thalamus

Pons

Medulla oblongata

**Figure 39** Midsagittal section of the brain.

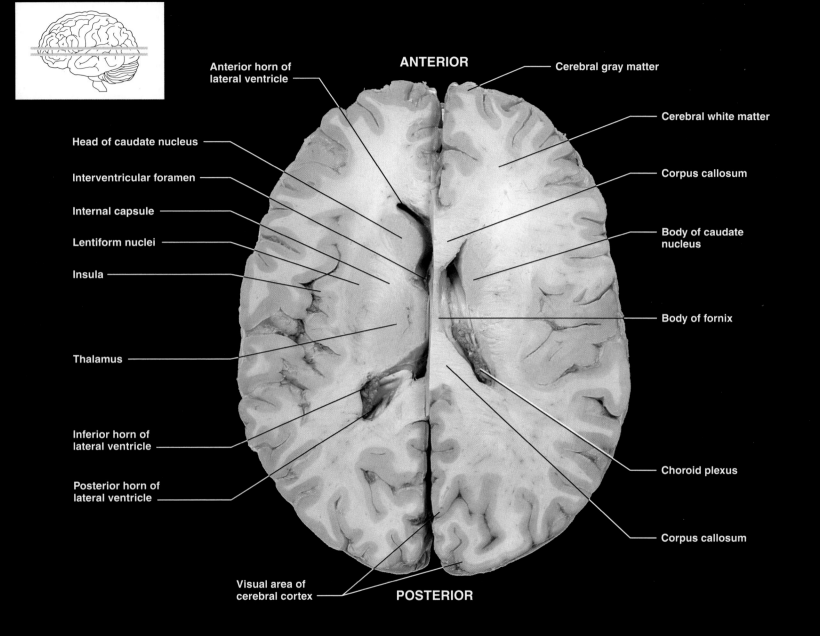

**Figure 40** Transverse section of the brain, superior view.
Left: on a level with the intraventricular foramen;
right: about 1.5 cm higher.

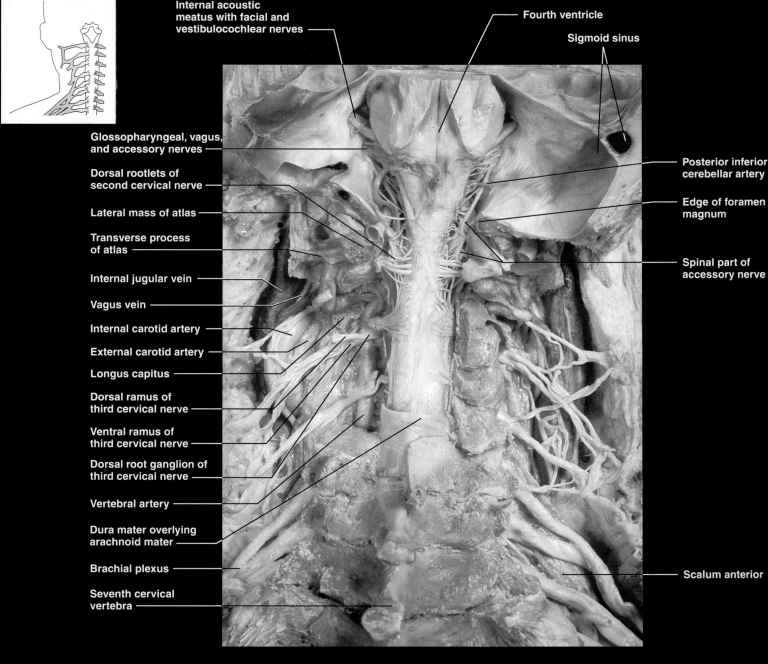

**Figure 41** Brainstem and cervical region of the spinal cord, posterior view.

Internal acoustic meatus with facial and vestibulocochlear nerves

Fourth ventricle

Sigmoid sinus

Glossopharyngeal, vagus, and accessory nerves

Dorsal rootlets of second cervical nerve

Lateral mass of atlas

Transverse process of atlas

Internal jugular vein

Vagus vein

Internal carotid artery

External carotid artery

Longus capitus

Dorsal ramus of third cervical nerve

Ventral ramus of third cervical nerve

Dorsal root ganglion of third cervical nerve

Vertebral artery

Dura mater overlying arachnoid mater

Brachial plexus

Seventh cervical vertebra

Posterior inferior cerebellar artery

Edge of foramen magnum

Spinal part of accessory nerve

Scalum anterior

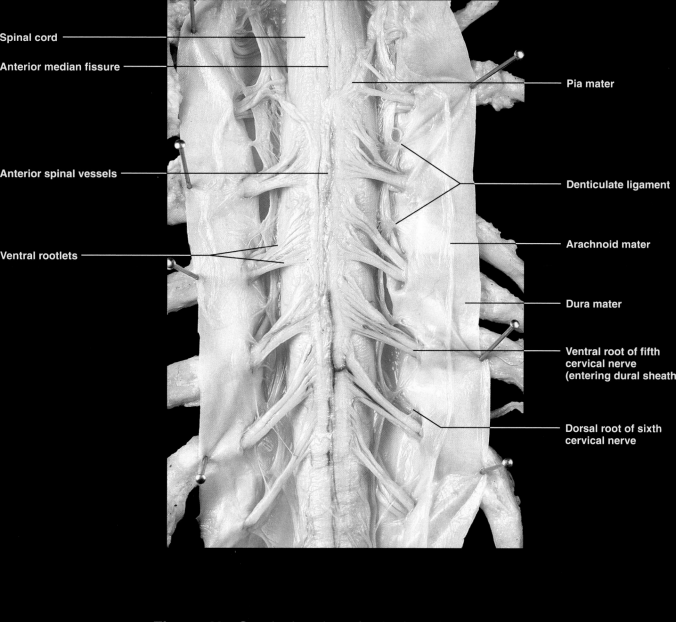

Spinal cord

Anterior median fissure

Anterior spinal vessels

Ventral rootlets

Pia mater

Denticulate ligament

Arachnoid mater

Dura mater

Ventral root of fifth cervical nerve (entering dural sheath

Dorsal root of sixth cervical nerve

**Figure 42** Cervical region of spinal cord, ventral view.

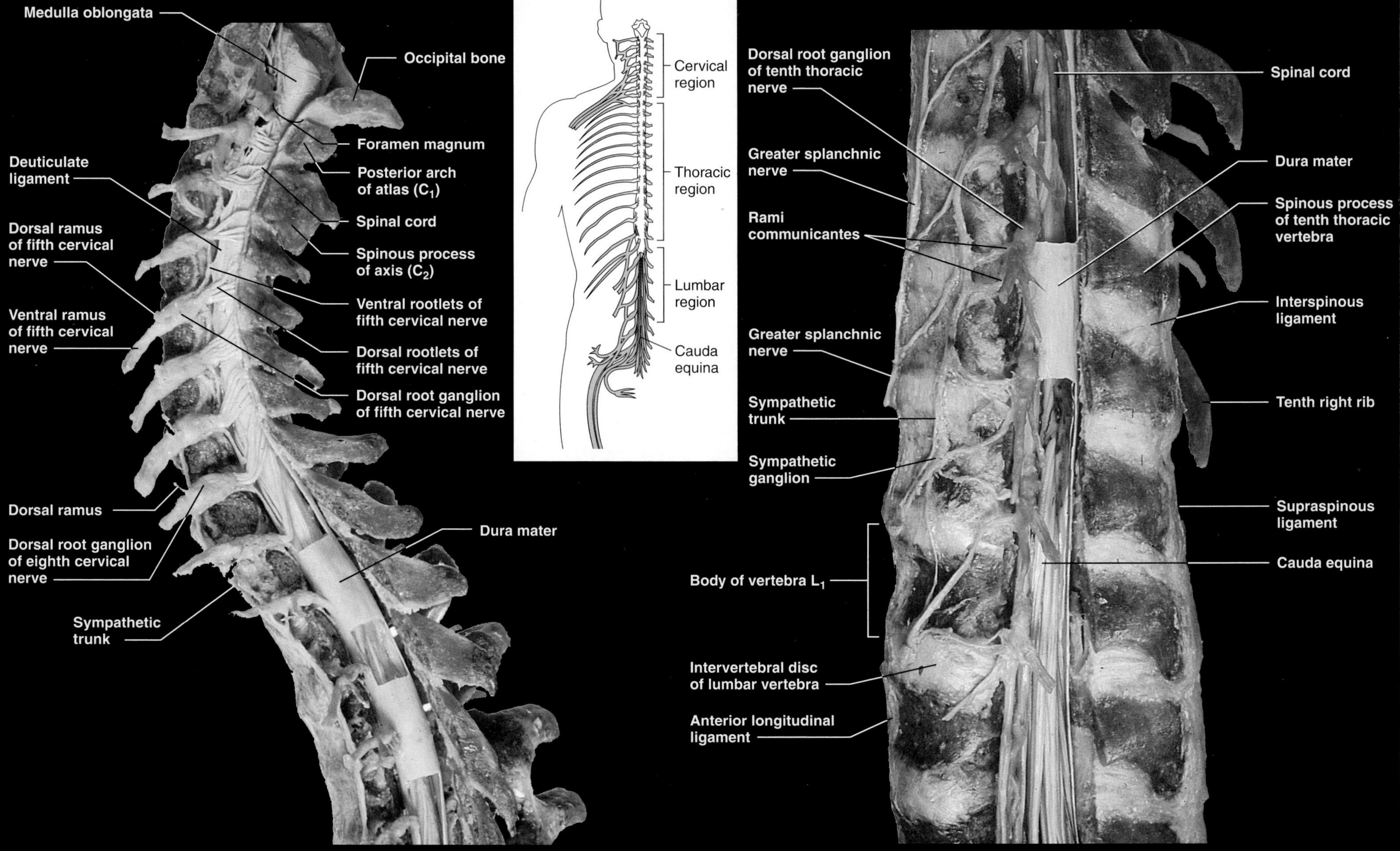

**(a)** cervical and upper thoracic regions from the left

**(b)** lower thoracic and upper lumbar regions from the left

**Figure 43** Vertebral column and spinal cord.

Labels for (a):
- Medulla oblongata
- Occipital bone
- Foramen magnum
- Posterior arch of atlas (C₁)
- Spinal cord
- Spinous process of axis (C₂)
- Ventral rootlets of fifth cervical nerve
- Dorsal rootlets of fifth cervical nerve
- Dorsal root ganglion of fifth cervical nerve
- Deuticulate ligament
- Dorsal ramus of fifth cervical nerve
- Ventral ramus of fifth cervical nerve
- Dorsal ramus
- Dorsal root ganglion of eighth cervical nerve
- Sympathetic trunk
- Dura mater

Inset labels:
- Cervical region
- Thoracic region
- Lumbar region
- Cauda equina

Labels for (b):
- Dorsal root ganglion of tenth thoracic nerve
- Greater splanchnic nerve
- Rami communicantes
- Greater splanchnic nerve
- Sympathetic trunk
- Sympathetic ganglion
- Body of vertebra L₁
- Intervertebral disc of lumbar vertebra
- Anterior longitudinal ligament
- Spinal cord
- Dura mater
- Spinous process of tenth thoracic vertebra
- Interspinous ligament
- Tenth right rib
- Supraspinous ligament
- Cauda equina

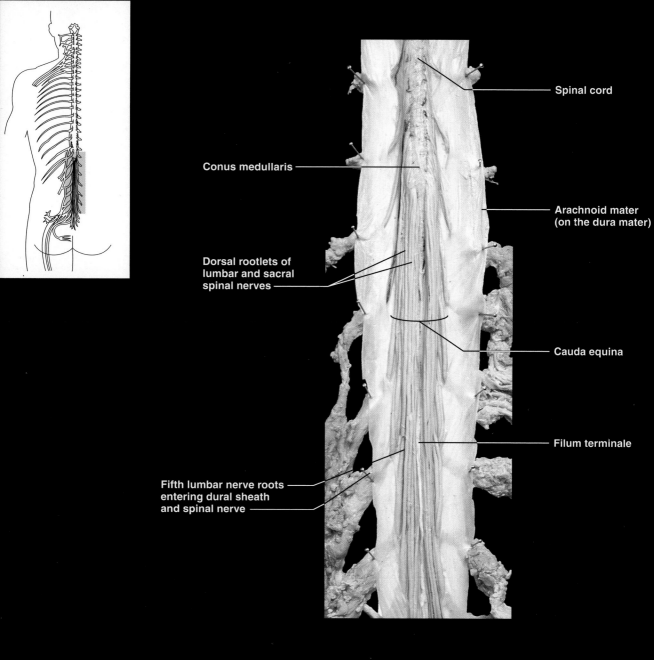

**Figure 44**  Spinal cord and cauda equina, dorsal view of lower end

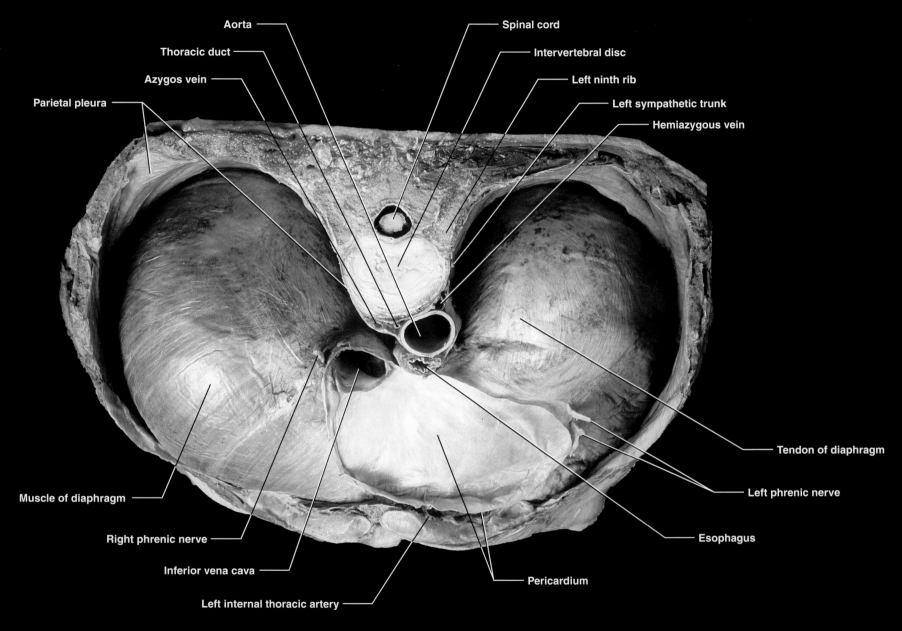

Aorta

Thoracic duct

Azygos vein

Parietal pleura

Spinal cord

Intervertebral disc

Left ninth rib

Left sympathetic trunk

Hemiazygous vein

Tendon of diaphragm

Left phrenic nerve

Esophagus

Muscle of diaphragm

Right phrenic nerve

Inferior vena cava

Left internal thoracic artery

Pericardium

**Figure 45** Diaphragm, superior view.

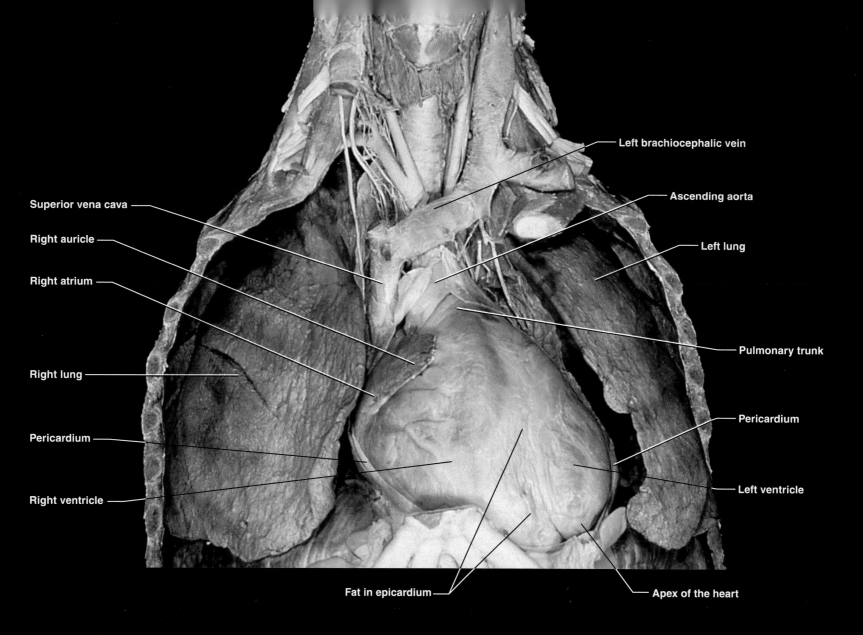

Left brachiocephalic vein

Superior vena cava

Ascending aorta

Right auricle

Left lung

Right atrium

Pulmonary trunk

Right lung

Pericardium

Pericardium

Right ventricle

Left ventricle

Fat in epicardium

Apex of the heart

**Figure 46** Heart and associated structures in thorax.

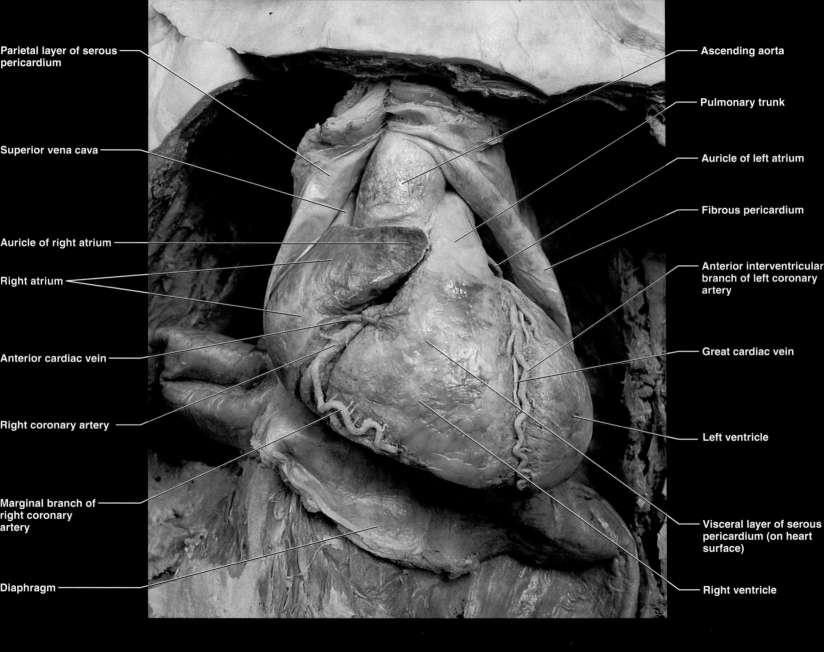

Parietal layer of serous pericardium

Superior vena cava

Auricle of right atrium

Right atrium

Anterior cardiac vein

Right coronary artery

Marginal branch of right coronary artery

Diaphragm

Ascending aorta

Pulmonary trunk

Auricle of left atrium

Fibrous pericardium

Anterior interventricular branch of left coronary artery

Great cardiac vein

Left ventricle

Visceral layer of serous pericardium (on heart surface)

Right ventricle

**Figure 47** Heart and pericardium, anterior view.

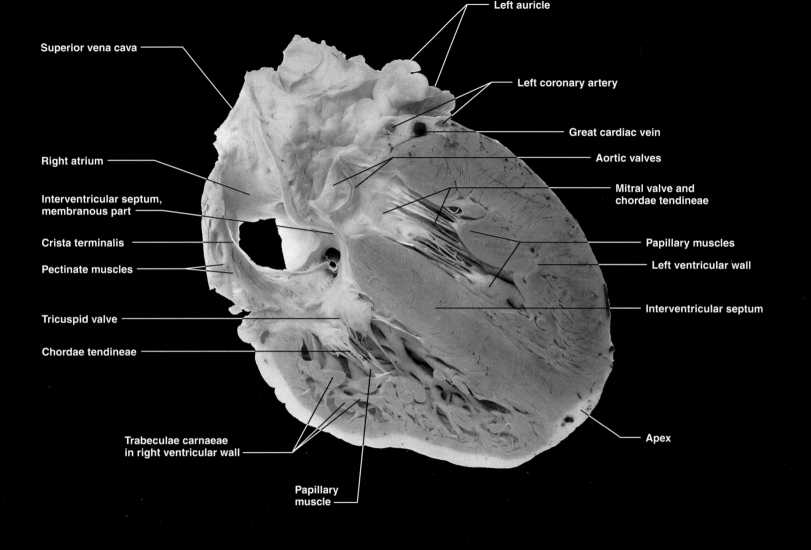

Left auricle

Superior vena cava

Left coronary artery

Great cardiac vein

Right atrium

Aortic valves

Interventricular septum,
membranous part

Mitral valve and
chordae tendineae

Crista terminalis

Papillary muscles

Pectinate muscles

Left ventricular wall

Tricuspid valve

Interventricular septum

Chordae tendineae

Trabeculae carnaeae
in right ventricular wall

Apex

Papillary
muscle

**Figure 48**  Coronal section of the ventricles, anterior view.

73

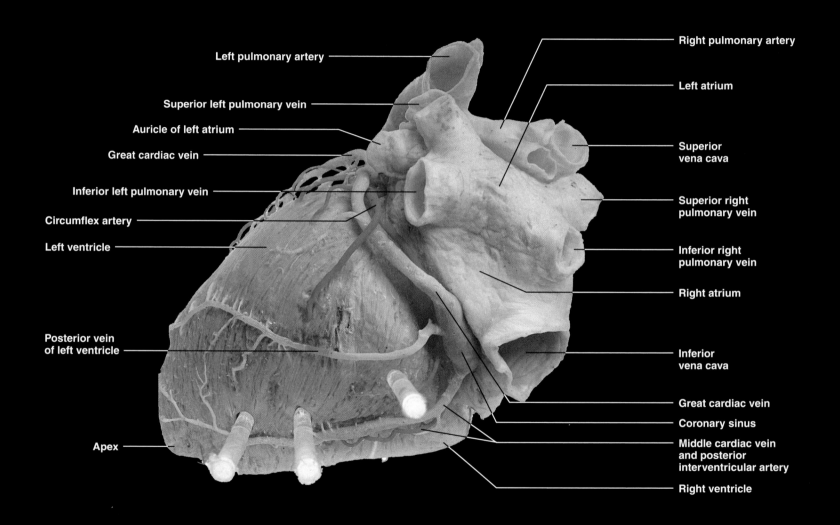

Left pulmonary artery

Superior left pulmonary vein

Auricle of left atrium

Great cardiac vein

Inferior left pulmonary vein

Circumflex artery

Left ventricle

Posterior vein
of left ventricle

Apex

Right pulmonary artery

Left atrium

Superior
vena cava

Superior right
pulmonary vein

Inferior right
pulmonary vein

Right atrium

Inferior
vena cava

Great cardiac vein

Coronary sinus

Middle cardiac vein
and posterior
interventricular artery

Right ventricle

**Figure 49** Heart, posterior view (blood vessels injected).

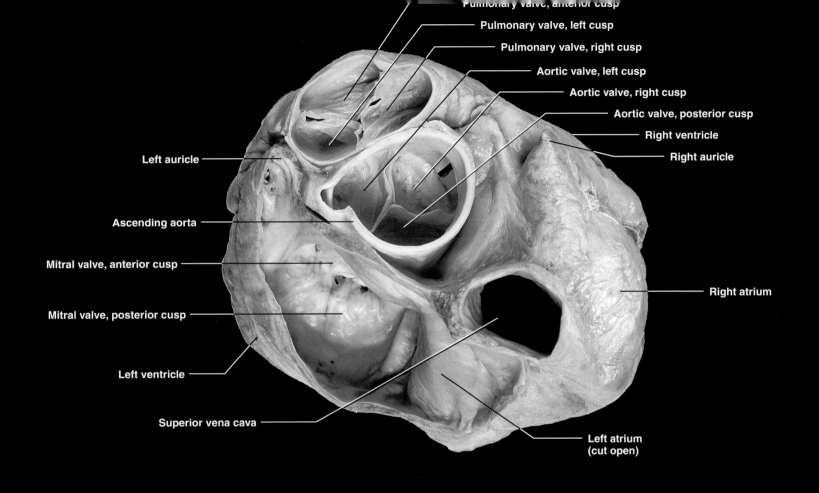

Pulmonary valve, anterior cusp

Pulmonary valve, left cusp

Pulmonary valve, right cusp

Aortic valve, left cusp

Aortic valve, right cusp

Aortic valve, posterior cusp

Right ventricle

Right auricle

Left auricle

Ascending aorta

Mitral valve, anterior cusp

Mitral valve, posterior cusp

Right atrium

Left ventricle

Superior vena cava

Left atrium
(cut open)

**Figure 50** Pulmonary, aortic, and mitral valves of the heart, superior view.

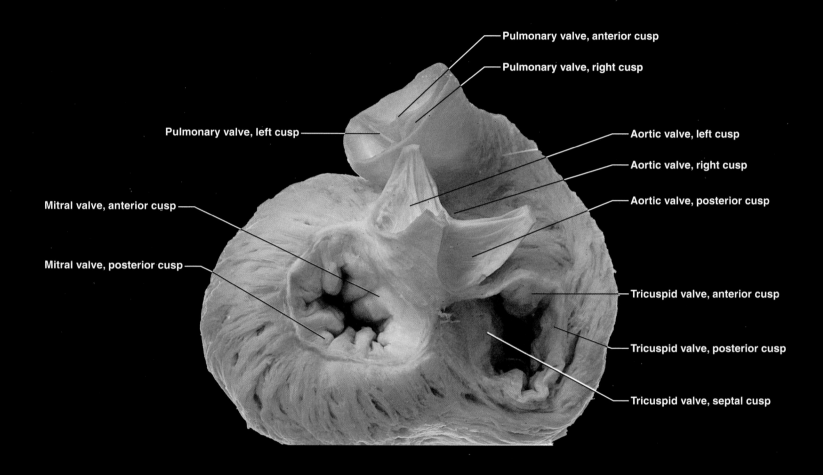

Pulmonary valve, anterior cusp

Pulmonary valve, right cusp

Pulmonary valve, left cusp

Aortic valve, left cusp

Aortic valve, right cusp

Mitral valve, anterior cusp

Aortic valve, posterior cusp

Mitral valve, posterior cusp

Tricuspid valve, anterior cusp

Tricuspid valve, posterior cusp

Tricuspid valve, septal cusp

**Figure 51**   Fibrous framework of the heart (atria removed), posterior view, from the right.

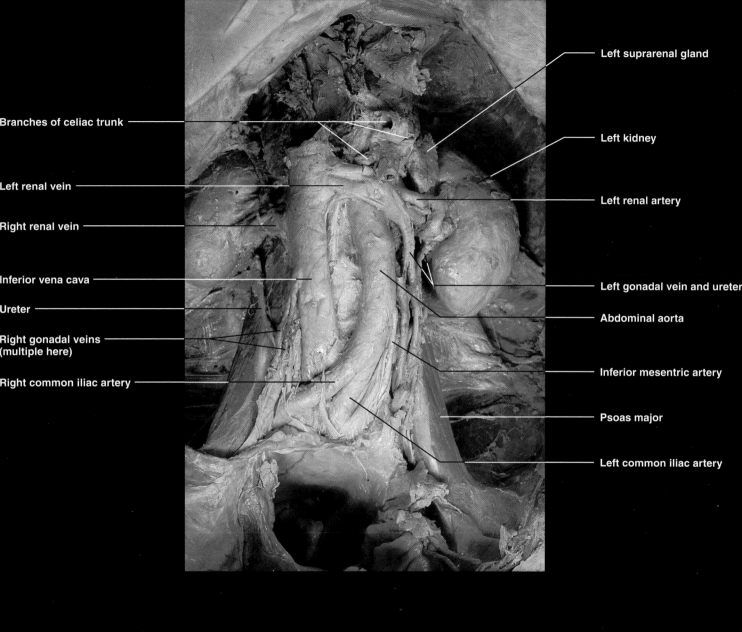

Branches of celiac trunk

Left renal vein

Right renal vein

Inferior vena cava

Ureter

Right gonadal veins (multiple here)

Right common iliac artery

Left suprarenal gland

Left kidney

Left renal artery

Left gonadal vein and ureter

Abdominal aorta

Inferior mesentric artery

Psoas major

Left common iliac artery

**Figure 52** Arteries in the abdominal cavity showing branches of the aorta.

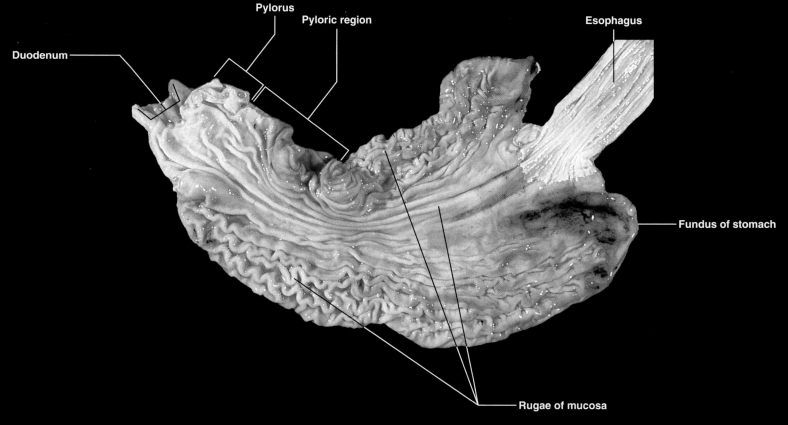

Pylorus

Pyloric region

Esophagus

Duodenum

Fundus of stomach

Rugae of mucosa

**(a)** frontal section of the internal surface of the stomach.

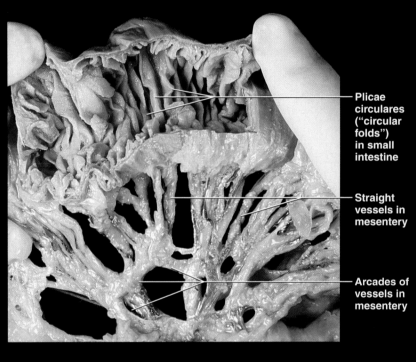

Plicae
circulares
("circular
folds")
in small
intestine

Straight
vessels in
mesentery

Arcades of
vessels in
mesentery

**(b)** small intestine, cut open to show
plicae circulares

**Figure 53** Internal surfaces of the stomach and small intestine.

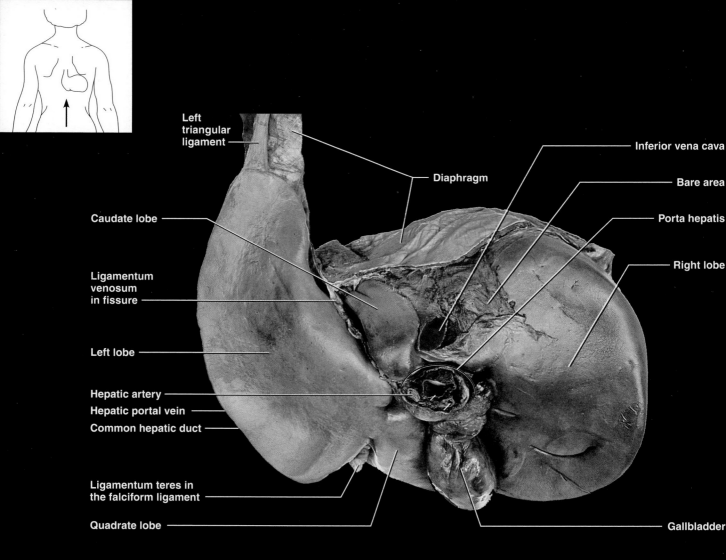

**Left triangular ligament**

**Diaphragm**

**Inferior vena cava**

**Bare area**

**Porta hepatis**

**Caudate lobe**

**Ligamentum venosum in fissure**

**Right lobe**

**Left lobe**

**Hepatic artery**
**Hepatic portal vein**
**Common hepatic duct**

**Ligamentum teres in the falciform ligament**

**Quadrate lobe**

**Gallbladder**

**Figure 54** Liver, posteroinferior view.

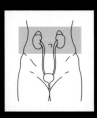

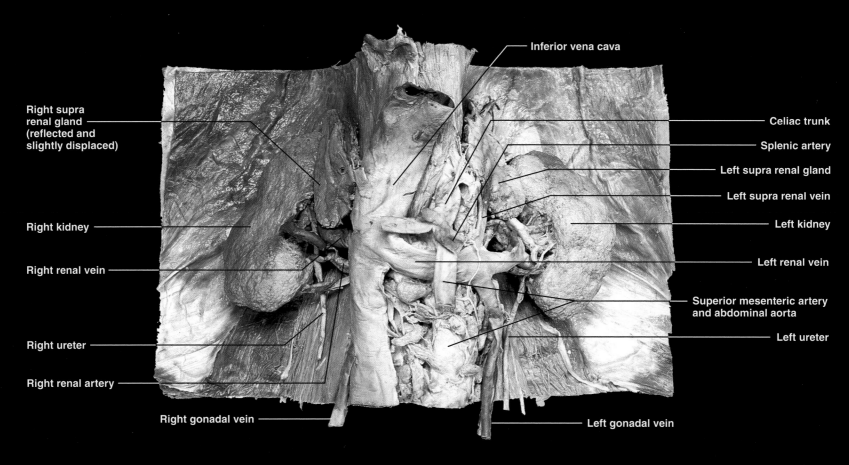

Right supra
renal gland
(reflected and
slightly displaced)

Right kidney

Right renal vein

Right ureter

Right renal artery

Right gonadal vein

Inferior vena cava

Celiac trunk

Splenic artery

Left supra renal gland

Left supra renal vein

Left kidney

Left renal vein

Superior mesenteric artery
and abdominal aorta

Left ureter

Left gonadal vein

**Figure 55** Kidney suprarenal gland and related vessels, anterior view.

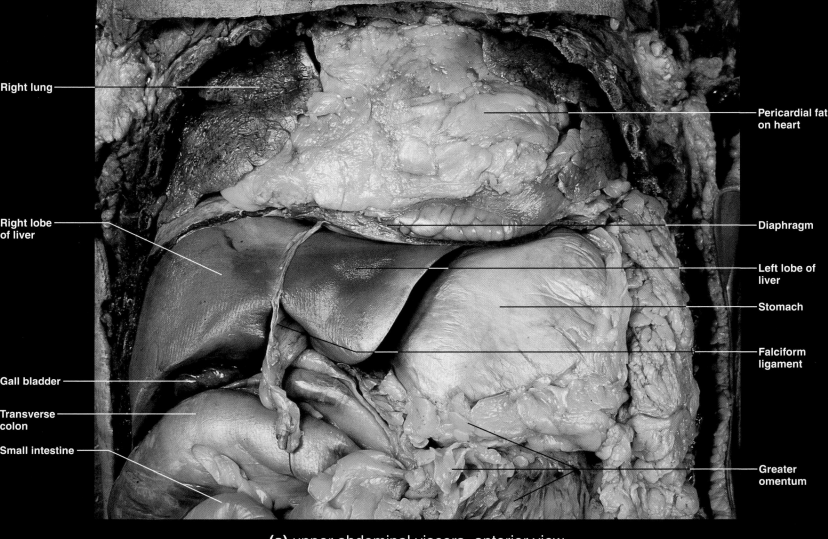

Right lung

Right lobe
of liver

Gall bladder

Transverse
colon

Small intestine

Pericardial fat
on heart

Diaphragm

Left lobe of
liver

Stomach

Falciform
ligament

Greater
omentum

**(a)** upper abdominal viscera, anterior view

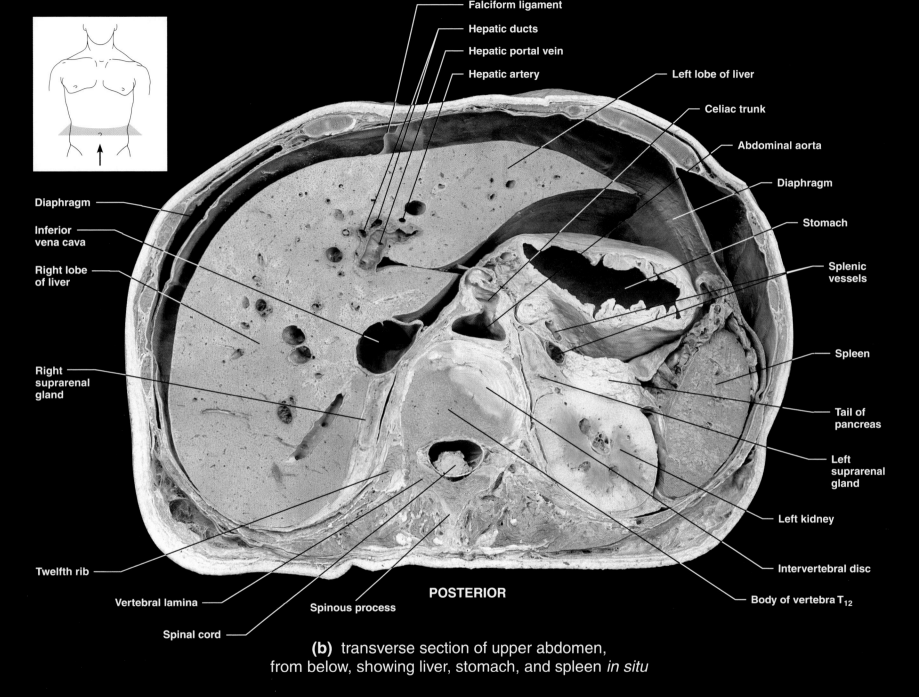

Falciform ligament

Hepatic ducts

Hepatic portal vein

Hepatic artery

Left lobe of liver

Celiac trunk

Abdominal aorta

Diaphragm

Stomach

Diaphragm

Inferior
vena cava

Splenic
vessels

Right lobe
of liver

Right
suprarenal
gland

Spleen

Tail of
pancreas

Left
suprarenal
gland

Left kidney

Intervertebral disc

Twelfth rib

**POSTERIOR**

Body of vertebra T$_{12}$

Vertebral lamina

Spinous process

Spinal cord

**(b)** transverse section of upper abdomen,
from below, showing liver, stomach, and spleen *in situ*

**Figure 56** Upper abdomen.

83

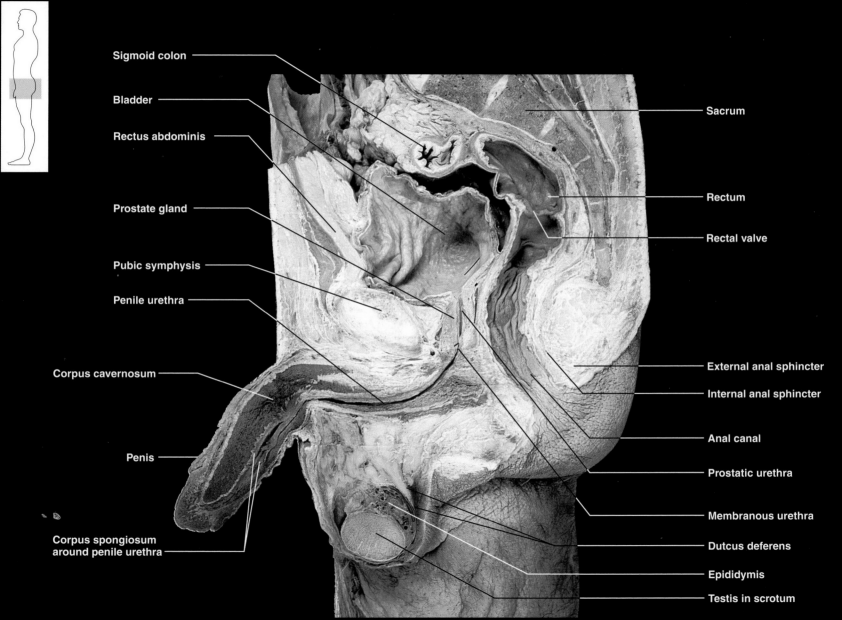

Sigmoid colon

Bladder

Rectus abdominis

Prostate gland

Pubic symphysis

Penile urethra

Corpus cavernosum

Penis

Corpus spongiosum
around penile urethra

Sacrum

Rectum

Rectal valve

External anal sphincter

Internal anal sphincter

Anal canal

Prostatic urethra

Membranous urethra

Dutcus deferens

Epididymis

Testis in scrotum

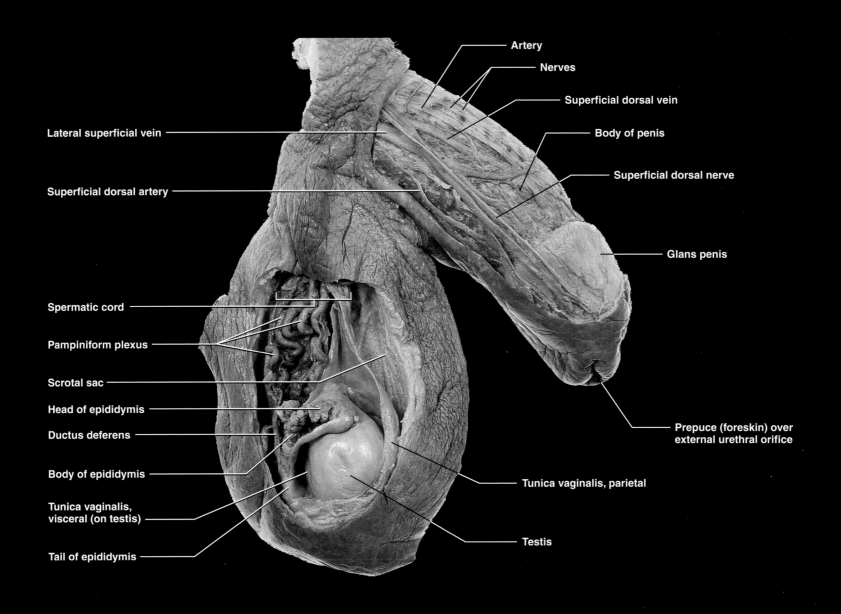

Artery

Nerves

Superficial dorsal vein

Lateral superficial vein

Body of penis

Superficial dorsal nerve

Superficial dorsal artery

Glans penis

Spermatic cord

Pampiniform plexus

Scrotal sac

Head of epididymis

Ductus deferens

Prepuce (foreskin) over
external urethral orifice

Body of epididymis

Tunica vaginalis, parietal

Tunica vaginalis,
visceral (on testis)

Testis

Tail of epididymis

**Figure 58** Section through the right testis and epididymis, and the penis.

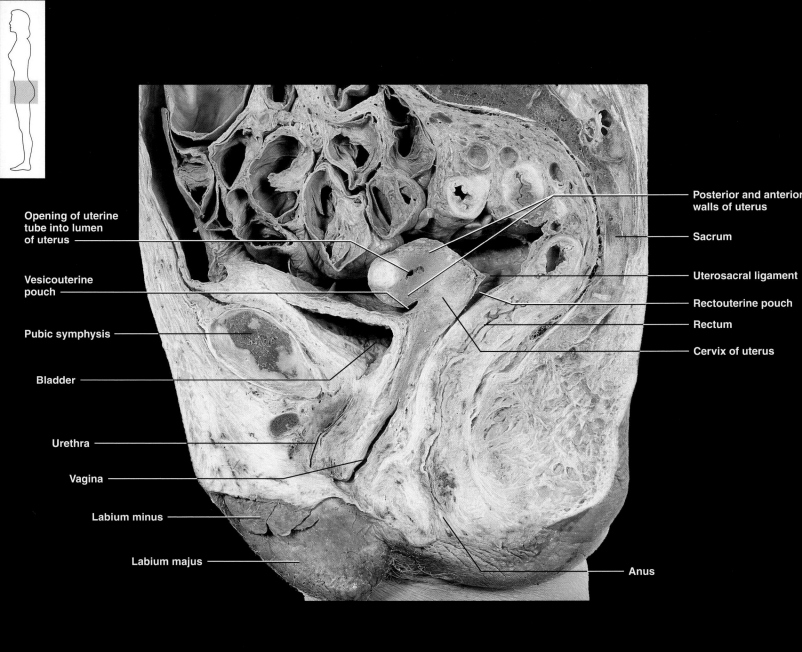

**Opening of uterine tube into lumen of uterus**

**Vesicouterine pouch**

**Pubic symphysis**

**Bladder**

**Urethra**

**Vagina**

**Labium minus**

**Labium majus**

**Posterior and anterior walls of uterus**

**Sacrum**

**Uterosacral ligament**

**Rectouterine pouch**

**Rectum**

**Cervix of uterus**

**Anus**

**Figure 59** Female pelvis, sagittal section (uterus points forward in this view).

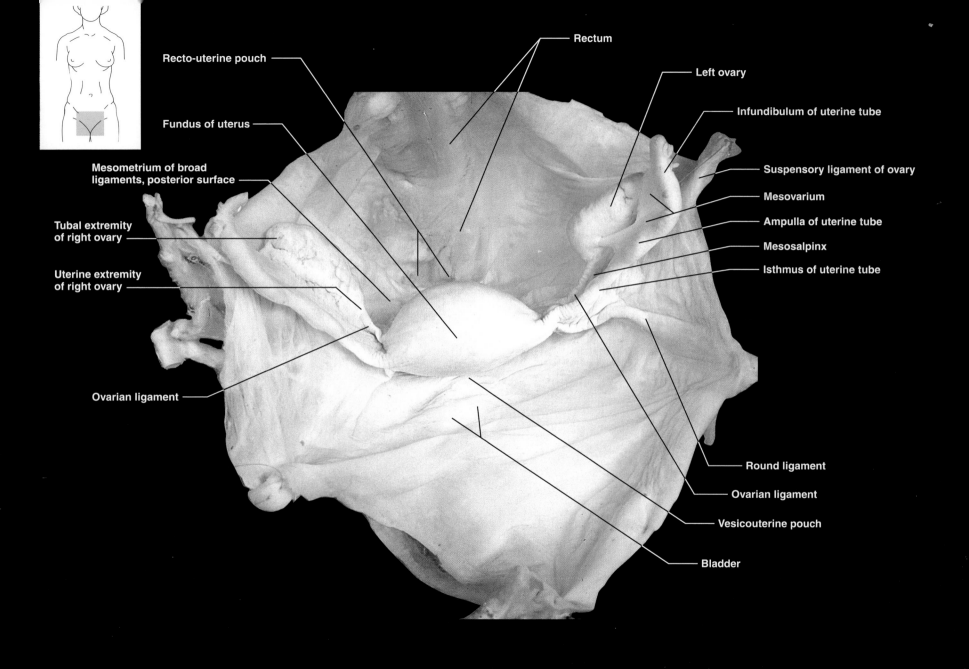

Rectum

Recto-uterine pouch

Left ovary

Infundibulum of uterine tube

Fundus of uterus

Suspensory ligament of ovary

Mesometrium of broad ligaments, posterior surface

Mesovarium

Ampulla of uterine tube

Tubal extremity of right ovary

Mesosalpinx

Isthmus of uterine tube

Uterine extremity of right ovary

Round ligament

Ovarian ligament

Ovarian ligament

Vesicouterine pouch

Bladder

**Figure 60**  Female pelvic cavity showing the position of the uterus relative to other structures.

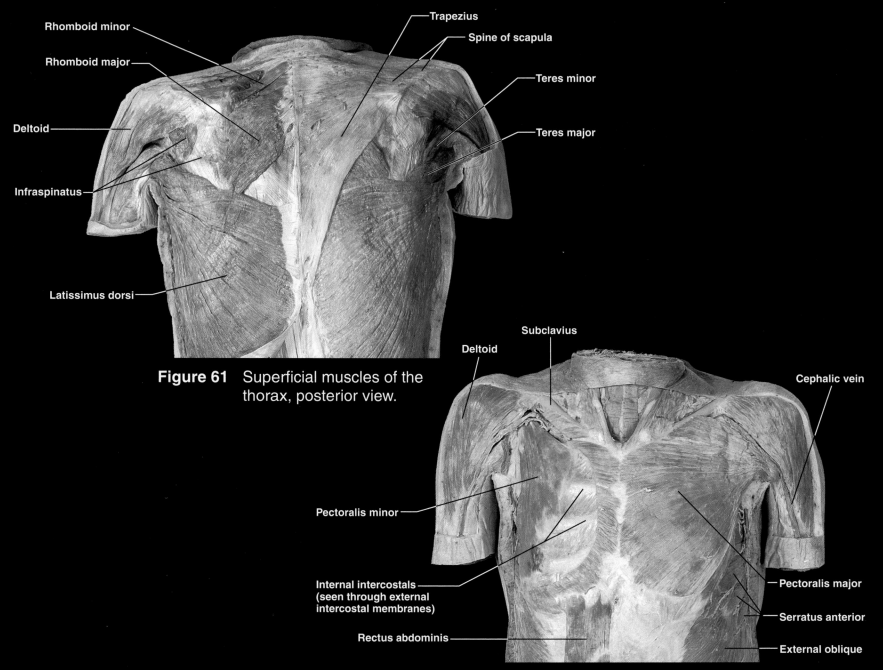

Rhomboid minor

Rhomboid major

Deltoid

Infraspinatus

Latissimus dorsi

Trapezius

Spine of scapula

Teres minor

Teres major

**Figure 61** Superficial muscles of the thorax, posterior view.

Deltoid

Subclavius

Cephalic vein

Pectoralis minor

Internal intercostals (seen through external intercostal membranes)

Rectus abdominis

Pectoralis major

Serratus anterior

External oblique

**Figure 62** Superficial muscles of the thorax, anterior view.

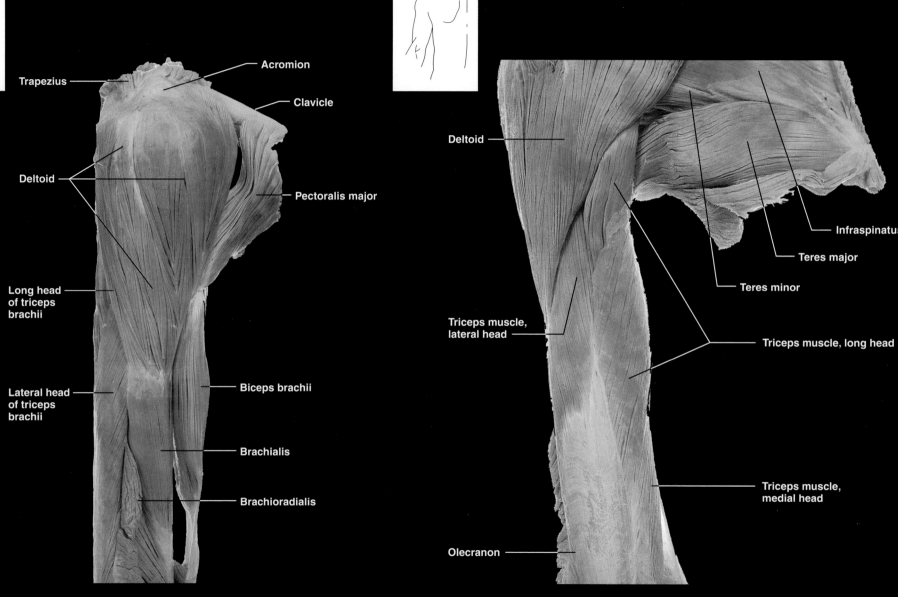

**Figure 63** Right shoulder from right, showing deltoid muscle and biceps.

**Figure 64** Triceps of the left arm, posterior view.

Trapezius

Acromion

Clavicle

Deltoid

Pectoralis major

Long head of triceps brachii

Lateral head of triceps brachii

Biceps brachii

Brachialis

Brachioradialis

Deltoid

Infraspinatus

Teres major

Teres minor

Triceps muscle, lateral head

Triceps muscle, long head

Triceps muscle, medial head

Olecranon

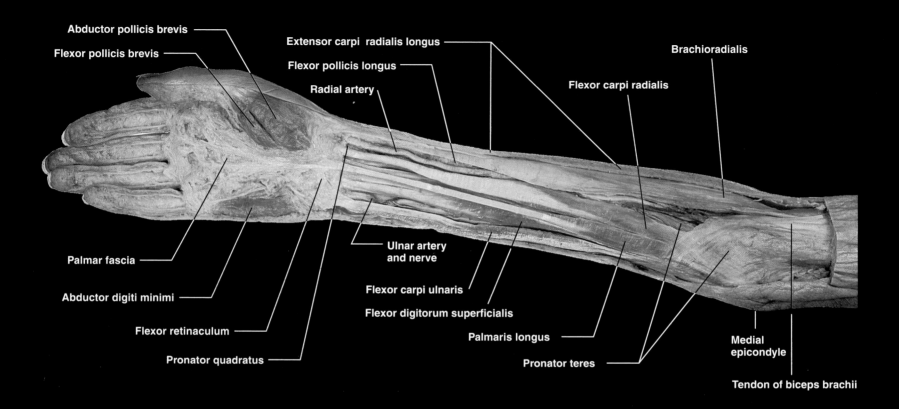

Abductor pollicis brevis

Flexor pollicis brevis

Extensor carpi radialis longus

Flexor pollicis longus

Radial artery

Brachioradialis

Flexor carpi radialis

Palmar fascia

Abductor digiti minimi

Flexor retinaculum

Pronator quadratus

Ulnar artery
and nerve

Flexor carpi ulnaris

Flexor digitorum superficialis

Palmaris longus

Pronator teres

Medial
epicondyle

Tendon of biceps brachii

**(a)** palmar surface

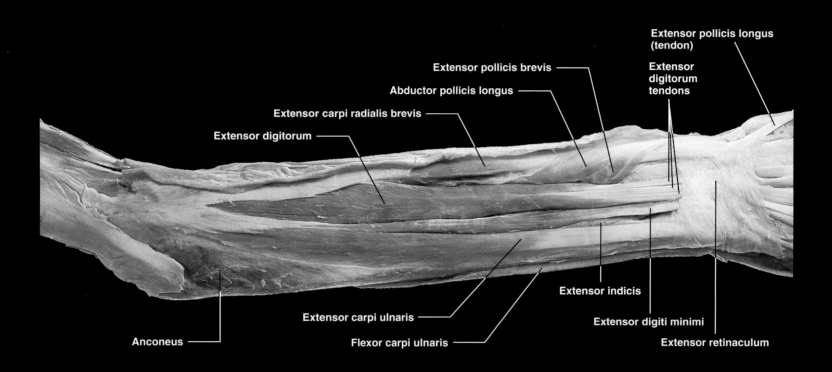

Extensor pollicis longus
(tendon)

Extensor pollicis brevis

Extensor
digitorum
tendons

Abductor pollicis longus

Extensor carpi radialis brevis

Extensor digitorum

Extensor indicis

Extensor carpi ulnaris

Extensor digiti minimi

Anconeus

Flexor carpi ulnaris

Extensor retinaculum

**(b)** dorsum surface

**Figure 65** Forearm and wrist.

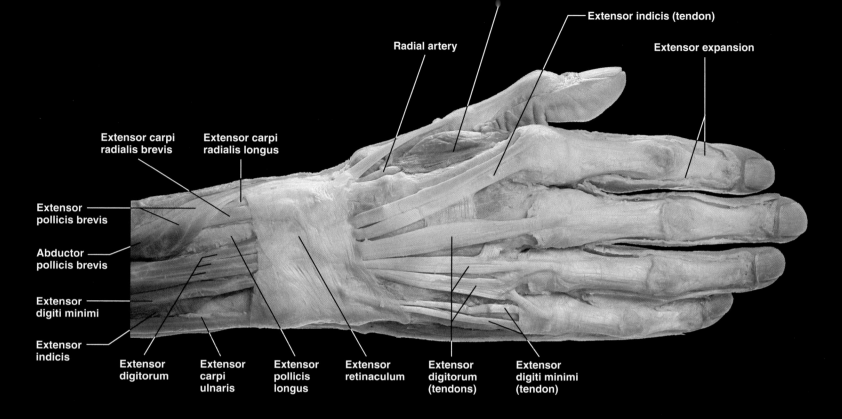

Extensor indicis (tendon)

Radial artery

Extensor expansion

Extensor carpi radialis brevis

Extensor carpi radialis longus

Extensor pollicis brevis

Abductor pollicis brevis

Extensor digiti minimi

Extensor indicis

Extensor digitorum

Extensor carpi ulnaris

Extensor pollicis longus

Extensor retinaculum

Extensor digitorum (tendons)

Extensor digiti minimi (tendon)

**(a)** dorsum surface of the right hand and wrist

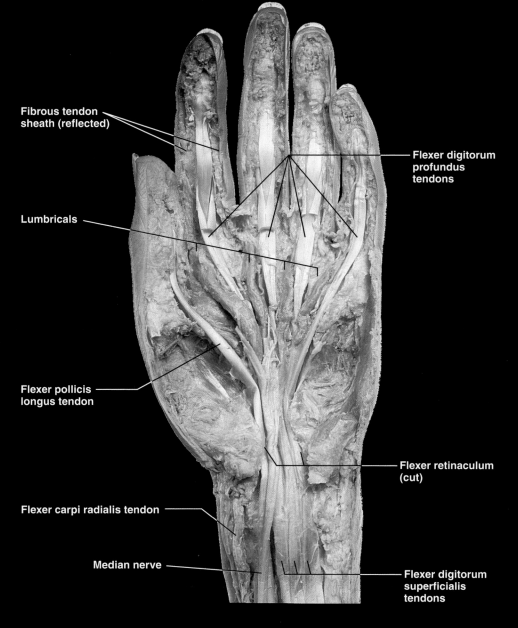

Fibrous tendon
sheath (reflected)

Flexer digitorum
profundus
tendons

Lumbricals

Flexer pollicis
longus tendon

Flexer retinaculum
(cut)

Flexer carpi radialis tendon

Median nerve

Flexer digitorum
superficialis
tendons

**(b)** palmar surface of the left
hand and wrist

**Figure 66** Wrist and hand.

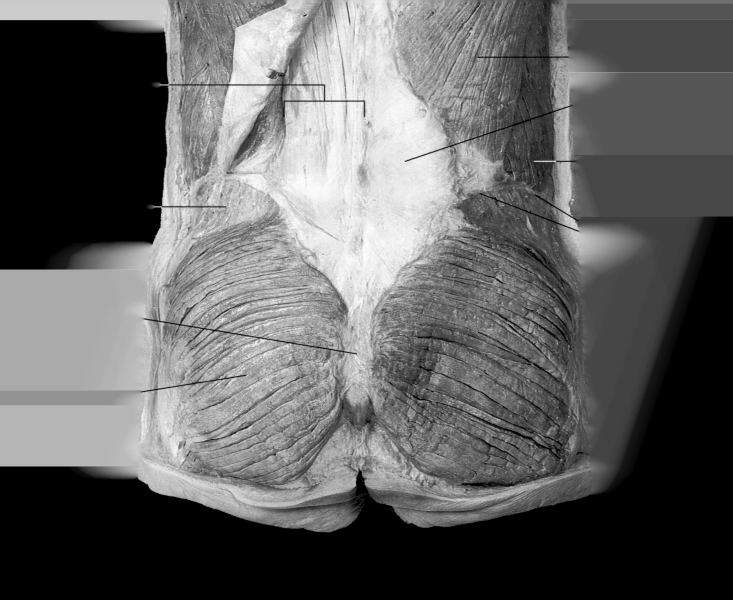

**Figure 67** Superficial muscles of the upper gluteal region.

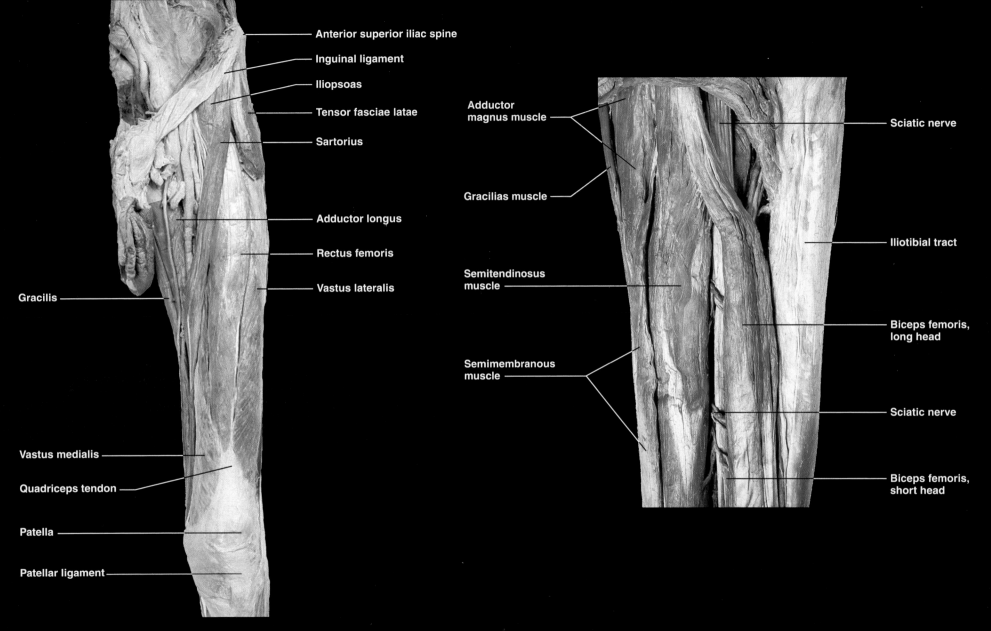

Anterior superior iliac spine

Inguinal ligament

Iliopsoas

Tensor fasciae latae

Sartorius

Adductor longus

Rectus femoris

Vastus lateralis

Gracilis

Vastus medialis

Quadriceps tendon

Patella

Patellar ligament

Adductor magnus muscle

Gracilias muscle

Semitendinosus muscle

Semimembranous muscle

Sciatic nerve

Iliotibial tract

Biceps femoris, long head

Sciatic nerve

Biceps femoris, short head

**Figure 68** Superficial muscles of the left lower thigh, anterior view.

**Figure 69** Right upper thigh, posterior view (the gap between the semitendinous and biceps opened to show sciatic trunk).

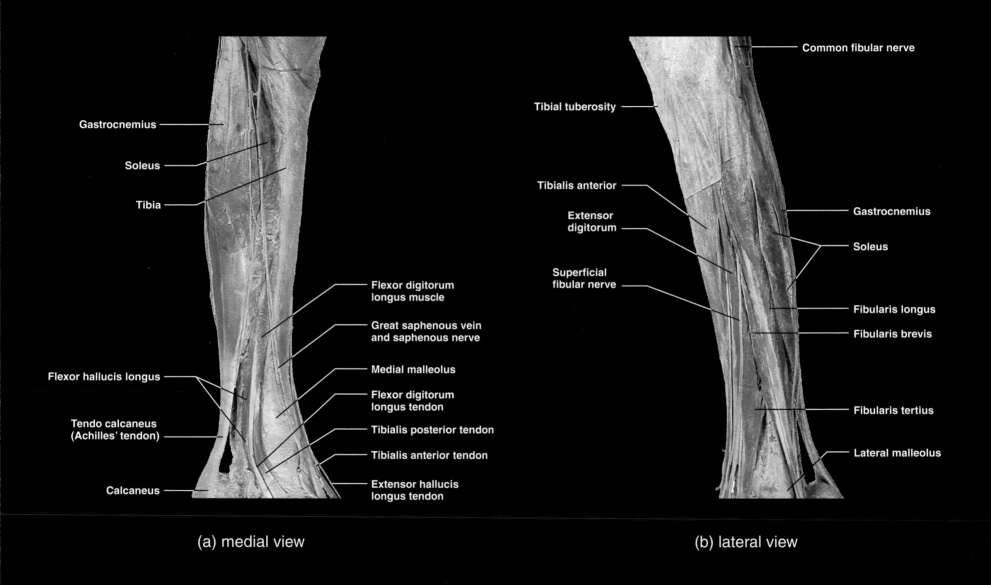

Gastrocnemius

Soleus

Tibia

Flexor hallucis longus

Tendo calcaneus
(Achilles' tendon)

Calcaneus

Flexor digitorum
longus muscle

Great saphenous vein
and saphenous nerve

Medial malleolus

Flexor digitorum
longus tendon

Tibialis posterior tendon

Tibialis anterior tendon

Extensor hallucis
longus tendon

Common fibular nerve

Tibial tuberosity

Tibialis anterior

Extensor
digitorum

Superficial
fibular nerve

Gastrocnemius

Soleus

Fibularis longus

Fibularis brevis

Fibularis tertius

Lateral malleolus

(a) medial view

(b) lateral view

**Figure 70**  Leg.

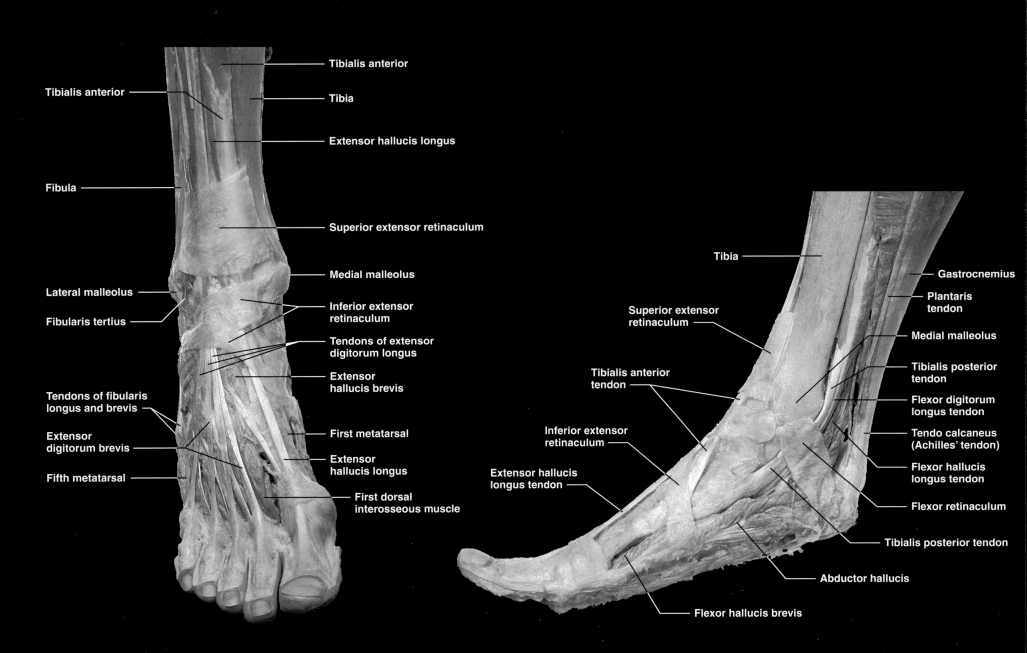

Tibialis anterior

Tibialis anterior

Tibia

Extensor hallucis longus

Fibula

Superior extensor retinaculum

Lateral malleolus

Medial malleolus

Fibularis tertius

Inferior extensor retinaculum

Tendons of extensor digitorum longus

Extensor hallucis brevis

Tendons of fibularis longus and brevis

First metatarsal

Extensor digitorum brevis

Extensor hallucis longus

Fifth metatarsal

First dorsal interosseous muscle

(a) anterior view

Tibia

Gastrocnemius

Plantaris tendon

Superior extensor retinaculum

Medial malleolus

Tibialis anterior tendon

Tibialis posterior tendon

Flexor digitorum longus tendon

Inferior extensor retinaculum

Tendo calcaneus (Achilles' tendon)

Extensor hallucis longus tendon

Flexor hallucis longus tendon

Flexor retinaculum

Tibialis posterior tendon

Abductor hallucis

Flexor hallucis brevis

(b) medial view